WHAT COMES AFTER HOMO SAPIENS?

WHAT COMES AFTER HOMO SAPIENS?

When and How Our Species Will Evolve into Another Species

DON SIMBORG

DWS PUBLISHING

What Comes After *Homo Sapiens*?

DWS Publishing
Mill Valley, California
dsimborg@gmail.com

Publisher's Cataloging-In-Publication Data

Simborg, Don.

What comes after *Homo sapiens*? : when and how our species will evolve into another species / Don Simborg.

First edition. | Mill Valley, California : DWS Publishing, 2017. | Includes bibliographical references and index.

ISBN 978-0-692-94603-9 (paperback) | ISBN 978-0-692-92001-5 (hardcover) | ISBN 978-0-692-94604-6 (ebook)

LCSH: Human evolution--Research. | Hominids--Speciation. | Evolution (Biology) | Genetic engineering. | Artificial intelligence.

LCC GN281 .S56 2017 (print) | LCC GN281 (ebook) | DDC 599.93/8--dc23

Project management: The Peter Beren Agency, PETERBEREN.COM
Book design: Mark Shepard, SHEPGRAPHICS.COM

First Edition published September 2017

Contents

List of Figures

Appendices Figures

Introduction

"My intent is to write not a science fiction story but rather a new story based on science."

We're all here right now, living on this planet. But who's up next?

What comes after *Homo sapiens*?

That simple question naturally leads to several more. First, how will we know when *Homo sapiens* (Latin for "wise man") has evolved into some other species (let's call that *Homo nouveau*)? Further, when will that happen, and what will that new species be like?

And, since none of us may be around when this new species arrives, why should it matter to any of us?

As you'll see as we move through the discussion in this book, it will matter a great deal. Further, it may be unlike any shift in species that came before it.

Let's start by looking at an evolutionary image—one I'm certain you've seen in some form—that shows the progression of humans from apelike creatures to our current state (figure 1).

FIGURE 1

This book will attempt to fill in the partially obscured figure at the end of this progression—the one just to the right of the *Homo sapiens*.

The *Homo sapiens* that looks just like every one of us.

Why should we take an interest in this? For one thing, we're the only species in the 3.8-billion-year history of life on earth that is even capable of doing that! We alone are equipped to ask those kinds of questions, let alone suggest possible answers. After all, the very concept of species didn't exist until *Homo sapiens* defined it.

We're also the only ones who have the tools necessary to effectively address those kinds of questions. No other species understands evolution, genetics and the mechanisms of speciation, among other concepts.

Perhaps most exciting of all, if and when *Homo nouveau* emerges, we will be the first species ever to be able to recognize that we have evolved into another one. That will be a singular event in the vast continuum of evolution.

But this also raises many puzzling questions:

- Who precisely will discover this? How will it occur? Will it be a sudden insight by some future taxonomist or will it come through careful study of fossil and genomic records?
- Will *Homo sapiens* still, in fact, exist when this recognition occurs? Will there be two *Homo* species coexisting? (If so, it won't be for the first time.)
- Just who will do the recognizing: *Homo sapiens* or *Homo nouveau*?
- How will we get along? Will we get along?

Tack on some pertinent questions of your own and you can start to see what a significant and provocative topic we're about to explore.

The Coming and Going of Species

It's helpful to back up at this point a bit and review some facts for greater perspective. We know that more than 99 percent of all species that once existed no longer do.[1] Species come and go—a seriously inconvenient truth. Additionally, their arrival and departure aren't necessarily uniformly spaced. For instance, there have been relatively short periods when new species appeared in great numbers, like the Cambrian period 540 million years ago.[2] In this case, "short" is still measured in millions of years.

Likewise, there have been relatively "short" periods of mass extinctions, defined as the loss of at least 50 percent of the species. There have been five such mass extinctions, the most recent one being 66 million years ago, which wiped out the dinosaurs and most other species.[3] In fact, we may be undergoing a sixth mass extinction as we speak, which may be relevant to discussion later in this book.[4][5] On the other hand, we believe some species like some cyanobacteria—bacteria that live through photosynthesis—have been on earth for billions of years and have survived these massive changes relatively intact.[6]

Most species, however, came into being gradually over millions of years. Most have subsequently died out gradually over a similar amount of time through the process we call evolution.

So where do we—the species *Homo sapiens*—fit in? For one thing, we've only been around in our current form for about 200,000 years—a mere sliver of the time there has been life on earth. To illustrate this, stretch out your arms wide in each direction. The distance from the tip of your right middle finger to the tip of your left middle finger represents the total time of life on earth. If you trim either fingernail, you cut off the total time that *Homo sapiens* has been around—not very long! Further, chances are we won't be around very much longer, evolutionarily speaking.

The Face of Evolution

Since we're discussing the evolution of our species, let's get a handle on what that involves.

Evolution occurs by random genetic change interacting with the environment. We'll get into that in more detail later. But that broad statement leads to other more specific and compelling questions:

- Does it take just one genetic alteration to change a species—or thousands?
- How do we draw the line from one species to the next? Where's the exact point where one species changes into another?
- Does species change happen over a small number of generations or are there increments of species change over many generations on some kind of continuum?
- If things aren't so consistent, does that mean that it's somewhat arbitrary when species change actually occurs?

Of course, there's always the possibility that nature will toss aside some of the rules that governed evolution to this point. Maybe the transition from *Homo sapiens* to *Homo nouveau* will occur in a completely different manner. Maybe it will happen faster than it has in the past.

It's also possible that completely new forces will play a role in this evolutionary change. Will *Homo nouveau* be induced by genetic engineering? By computers? By other technology? By a global catastrophe? Lastly, give some thought to the most remarkable possibility of all—will we ourselves be the ones who create this new species? In effect, will *Homo sapiens* create *Homo nouveau*?

Those are all powerful questions. The objective of this book is to address them as comprehensively as possible.

That said, a great deal of science and technology will be discussed. But you won't need to be a scientist or particularly technologically savvy to understand and enjoy it.

On the other hand, I don't want to exclude scientists and technologists. Through this book, I hope to generate discussion on my website, whatcomesafterhomosapiens.com. Hopefully, the scientists and technologists whose work I have covered will comment on my conclusions and provide their own.

Since the overall scope of this topic is so enormous, I have by necessity summarized and synthesized much of the research I cite. To some scientists and technologists in those fields covered, this may seem superficial, incomplete and even misleading and incorrect. I welcome such feedback on my website. My intent is to write not a science fiction story but rather a new story based on science.

The Greatest Challenge of All

Another interesting element to this project is that there are really no experts on this particular subject. That is because it is not a single field of study, but rather many. In that respect, there's really no one who's "qualified" to write about this topic, at least by the traditional definition.

That's made this project an enormous challenge. But it's what has also made this project so exciting—an excitement I'm certain you'll come to share.

Here's why.

In his heavily researched book *Cosmosapiens: Human Evolution from the Origin of the Universe*, John Hands nicely summarizes the core challenge that this book addresses:

> *To be a successful scientist today means spending a career on colloid chemistry or palaeoarchaeology or studying chimpanzees, or some equally if not more specialized field. This narrowness of inquiry has produced a depth of knowledge but has also carved out canyons of expertise from which its practitioners find it difficult to engage in meaningful dialogue with other specialists except where canyons intersect in a cross-disciplinary study of the same narrow subject.*
>
> *Few scientists transcend their specialist field to address fundamental questions of human existence such as what are we? The few who do rarely engage in open-minded debate. Too often they are unable to see the bigger picture and from their canyons they tend to fire a fusillade of views derived from the training, focus, and culture of the narrow academic discipline in which they have spent their professional lives.*[7]

Think about that for a moment. Eminent scientists, researchers and others possess extraordinarily levels of knowledge and insight, yet rarely piece that together with other knowledge to address even greater questions. In a sense, the specific focus of their impressive credentials effectively creates a barrier that cripples their ability to gain greater perspective.

So, who then can see what these accomplished scholars often cannot? You and me, for starters. Let's attempt to cross those canyons.

My Preparation for Writing this Book

With regard to myself, my background in medicine and scientific research has prepared me well to undertake a project like this. I've published more than 100 peer-reviewed articles and have served on the faculty in the schools of medicine at Johns Hopkins and the University of California, San Francisco. My training and practice were in internal medicine, and my specialty is medical informatics—using computers for electronic medical records, diagnosis and other purposes. I founded and led two companies focused on clinical decision support and electronic medical records.

Although I have a strong background in medicine and scientific research, I'm not an expert in any of the fields this book touches on (evolutionary biology, genomics, paleontology, speciation, artificial intelligence, neuroscience and many more). This has allowed me to analyze the pertinent data without any sort of biased or filtered eye.

That, I suspect, is the same sort of perspective you as the reader will also bring to this book—a fresh, untouched viewpoint that will make our shared process of investigation and discovery all the more compelling.

Because I took on this project with an unbiased perspective, this book was a true investigation on my part—a pure learning experience. This was not an attempt to justify some preconceived theory on my part. Each avenue I explored seemed plausible at first. Some paths led to likely threats to our very existence, like artificial intelligence, which I now believe to be our greatest existential threat (a surprise to me.) On the other hand, nanotechnology, the branch of technology focusing on dimensions of less than the width of a human hair, could be either an existential threat or the savior of humankind.

Surprises

There were a great many more surprises and unexpected paths to follow. Here are a few:

- I learned how brilliant and correct Darwin was for all the right reasons.
- I also learned how correct Jean-Baptiste Lamarck, the famous French biologist, was for all the wrong reasons.
- Gregor Mendel led us to an understanding of genetics that is dramatically oversimplified; his work was possibly falsified, yet Mendelian genetics is still foundational.
- Ray Kurzweil is predicting, "The singularity is near." Bill Joy is the cofounder of Sun Microsystems. They have opposite views of our future, but both are equally frightening.
- The genome—the genetic material of an organism—is so magnificent and complicated that it may take centuries for us to fully understand it.
- We have about the same number of genes as a mouse, but it is the epigenome—the part of the genome that regulates our genes—that really makes us human.
- I had never heard of the "species problem" before my research. I still don't understand it.
- We don't know for sure how and when *Homo sapiens* got here, but Svante Pääbo and his team are revolutionizing our understanding of human evolution.
- Our main differentiator is our brain, not our upright posture. We still don't know why the latter evolved.
- Our tools to study the brain are brilliant examples of the epitome of evolution, but our understanding of how the brain works is still in the future.
- CRISPR will take us closer to the Methuselarity. You will soon learn about both.

I'll save the discussion of these and other surprises for later in the book. The overriding point is that, unlike many scientists and researchers, my perspective was in no way limited by what I already knew or, perhaps all the more important, what I expected to find.

That's an advantage you enjoy as well. So, let's begin our journey.

Part 1

Getting to *Homo Sapiens*

Chapter 1

The Questions

"*Homo sapiens* will likely still exist when *Homo nouveau* emerges. That really makes it interesting!"

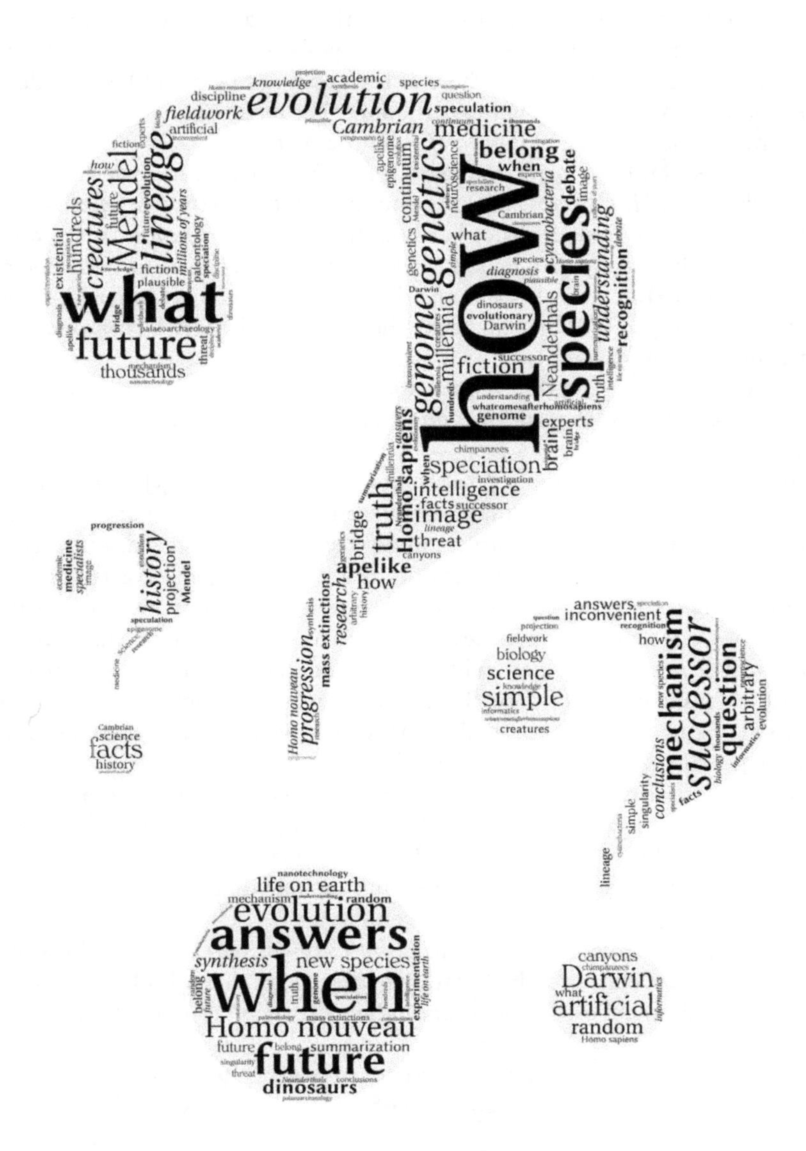

Lots of questions were raised in the introduction. I'm going to attempt to answer all of them. The main one of course is what comes after *Homo sapiens*? Before I begin discussing the answers, however, I need to clarify the questions and my process for getting to the answers; otherwise, the answers will be meaningless. Please resist the temptation to skip ahead to Chapter 14. You'll be so much better prepared to appreciate and understand the answers if you follow along with me. And it will be a lot more fun.

We are here today—the only species of humans left standing after a couple of million years since the earliest human species showed up on earth. There were many human species before *Homo sapiens*. I'll talk about the other humans later. But for now, I want to focus on the question of where we're heading—where we're evolving. Yes, we are still evolving. Sooner or later, we will evolve sufficiently to the point where at least some of us won't be considered *Homo sapiens* any more. What exactly does that mean? How will we know? When might that happen? How might that happen?

There are several implications of this. First, *Homo nouveau* will be a different species from *Homo sapiens*, so I examine what, exactly, a species is and how we will define the difference between the two. For example, there was once another human species called *Homo neanderthalensis*—or Neanderthals. They are now extinct. I'll describe the difference between Neanderthals and us in order to help project what might be the difference between *Homo sapiens* and a future *Homo nouveau.*

Second, I am assuming that *Homo nouveau* will still be considered a human. I'll elaborate on what that means and discuss what alternatives there might be for that.

One final implication is that *Homo sapiens* will still exist when *Homo nouveau* first arrives. That really makes it interesting! Therefore, I am not talking about events that happen on earth if and when *Homo sapiens* disappears from the earth. I am not trying to predict doomsday for our species or the earth in general; in fact, it is just the opposite. I'm presuming that *Homo sapiens* will survive long enough for our successor species to emerge. I'm saying that when *Homo nouveau* emerges, there will be two human species coexisting on earth at least for some period of time. That sounds weird, but actually that has been the usual circumstance in our evolutionary history. What may be unusual is how we get there, but that's for later.

If by some chance some catastrophe happens that wipes out *Homo sapiens*, as it did to the dinosaurs sixty-six million years ago, the question becomes moot. There will be no *Homo nouveau* in this case, as I discuss in chapter 10. There are other existential threats to *Homo sapiens* that could prevent us from evolving further, which I'll discuss as well.

Will *Homo nouveau* be superior to *Homo sapiens* in some way? Maybe. Maybe not. It depends on which path we take to get there. I'll leave that to the reader's judgment and provide enough information to make that call in chapter 14.

The question of describing the difference between *Homo sapiens* and *Homo nouveau* revolves around the definition of the word species. As shown in chapter 3 that is not as easy to define as one might expect. How different will they have to be? Who will decide? Lots of very interesting questions. We have many differences now within our population. Do we already have *Homo nouveau*?

I explore in chapter 4 how new species have emerged in the past, which is a process called speciation. I have considered also how speciation might be different going forward. In any case, species and speciation both require some understanding of genetics. That won't change, so I will dive into that in chapter 5. Hopefully the dive is not too deep. My goal is to provide just enough knowledge of genetics to allow understanding of the ensuing discussion. I moved a lot of material about genetics into appendix 4 (Everything You Didn't Want to Know about Genetics) if you want to learn more—but you can skip that and still understand the answers.

It was informative in this process to understand how we got to be *Homo sapiens* in the first place. Yes, figure 1 implies we came from the apes, but is that really true? Will the apes of today become humans one day? What exactly happened to get to humans? Chapters 6–8 will review what I learned about that and it was any eye opener to find out both what we know and don't know. That learning process helped me focus on the future pathways. That will take us to *Homo sapiens* and complete Part I of the book.

Then the more speculative part of the book begins. Part II will take us to *Homo nouveau.*

Will evolution for us continue more or less as it has in the past? In chapter 9 I'll describe the most likely alternatives. There are four major possibilities.

Chapters 10–13 will examine each of these possible pathways to *Homo nouveau.* Although the discussion of each is speculative, they are examined in reference to what is already known about the sciences of genetics, evolution and speciation.

This will lead to the answers in chapter 14.

Chapter 2
Taxonomy

"**D**id **K**ing **P**hilip **C**all **O**ut **F**or **G**ood **S**oup?"

I hated taxonomy when I studied it in college. It was boring and required a lot of memorization. So don't worry, we're just going to touch on it here. There are only a small number of taxonomic concepts that are important to this discussion.

The taxonomy of living things

King Philip Called Out For Good Soup. That's one of the many mnemonics people used when I studied it to remember the taxonomy of living things like plants and animals: kingdom, phylum, class, order, family, genus, species. It is a hierarchy of all living organisms from the most broad to the most specific. Most of those layers are not relevant to our discussion. Our focus is on the bottom of this massive classification system of all living things: species. You can't say what something living is without stating its species.

The great geneticist of the early twentieth century, Theodosius Dobzhansky, pointed out two things about taxonomy:

It is the great diversity of living organisms that led to the need for some type of classification system, and

It is the discontinuity of this diversity that enables a classification system.

That is, living and extinct diversity is not one big, single continuous array of characteristics. Instead, there are these gaps between groups of organisms that don't have clear intermediaries.[8] Without those gaps, it would be near impossible to develop the taxonomy.

Dobzhansky is quick to point out that evolutionary theory implies there have been gradual and continuous changes to organisms, and if we were able to assemble all of the organisms that ever existed, we would not have these gaps that enable our taxonomy. It is only because the intermediaries are extinct that we have them. That makes it a lot easier to distinguish closely related species. As you'll see in the next chapter, sometimes the gaps are very small and very debatable.

On the other hand, Stephen Jay Gould argued that the opposite might be true and that the gaps in the fossil record may be real reflections of evolution. He argued that much of the fossil record demonstrates long periods (many millions of years) where individual species do not change at all. This is coupled with the sudden appearance of a new species that is dramatically different from its predecessors. He referred to this as punctuated equilibrium, in contrast to the more widely accepted gradualism model of evolution.[9]

This debate continues to today and is very relevant to whether *Homo nouveau* will be a big leap from *Homo sapiens* or a barely perceptible modification.

(Every so often in these chapters I will highlight a key sentence as above that is a clue to the answers.)

Even our taxonomy is evolving.

I don't want to get into all the ins and outs of taxonomy because my focus is only on one component: species. Even more narrowly, we are focusing on species within the *Homo* genus. There are living things other than plants and animals, and it wasn't until after I got out of high school that the taxonomists decided that there was a classification level higher than kingdom, called domain. Bummer! I had to relearn my mnemonic to be Did King Philip Call Out For Good Soup. But again, this book is not about taxonomy; it's about our species, *Homo sapiens*.

Taxonomy of *Homo sapiens*
Domain: Eukarya
Kingdom: Animalia
Phylum: Chordata
(subphylum: Vertebrata)
Class: Mammalia
Order: Primate
Family: Hominidae
Genus: Homo
Species: Sapiens

However, as you will see, domain actually will become relevant regarding one possibility for *Homo nouveau*.

Taxonomy conventions (the boring part)

Note that the convention when referring to a species is to use two Latin terms, *italicized*, the first of which is capitalized and refers to the genus (in this case, *Homo*). The second refers to the species within the genus (in this case, *sapiens)*. I suppose in order to go along with convention, I should refer to *Homo nouveau* as *Homo aliusus* ("other human"), *Homo novus* ("new human"), or some other Latin name. The grandson of Charles Darwin, Charles Galton Darwin, suggested the name *Homo sapientior* ("wiser man") when speculating about a future *Homo* species. Of course, that would imply that the next species is wiser than the current one—something we don't know at this point. I'm going to stick with *Homo nouveau*.

As we'll see later, we're in a paradigm shift with regard to evolution, so our semantics can shift as well.

What exactly is Homo sapiens?

Carl Linnaeus, who originated this classification scheme in the mid eighteenth century, was a bit of a classification junkie. He wanted to fit just about everything, including non-living things, into neat and orderly groups. The box above shows this orderliness for *Homo sapiens* in the Linnaean version of taxonomy. What does that really mean? Simply put, we *Homo sapiens* have declared that there exists a species called *Homo sapiens*—and we are it.

When I say we are it, I mean everyone we think of today as human beings. It includes Hispanics, Caucasians, Africans, Asians, Inuits, Islanders, Aborigines, Native Americans, tall people, dwarves, people with genetic abnormalities, people born with one kidney, people with six fingers on each hand, people with an extra Y chromosome, and combinations of any of the above. We are all *Homo sapiens.*

I used the term *human beings*, but that is not really a technical term even though everyone knows what we mean by it. Technically, the word *human* refers to our genus, *Homo,* not our species. If there were Neanderthals or other species of the *Homo* genus still alive today, they would also be called humans or human beings. It turns out that the only surviving species of the genus *Homo* is *Homo sapiens*, so human beings that are alive today are all *Homo sapiens*. That is not true for all of history, of course, and it won't be true when *Homo nouveau* emerges. They will be humans.

I have refrained from using the term *the human race*. Similar to *human beings,* it generally refers to all *Homo sapiens*. However, the term *race* introduces the notion that *Homo sapiens* are often divided into groups that are called races. When other species are subdivided into groups, we refer to those groups as subspecies. Are human races subspecies of *Homo sapiens*? This raises a subject that is fraught with controversies, biases, historical misunderstandings and many other difficulties. I will attempt in later chapters to tease out the science from the myth associated with this topic, but for now, I will dodge this topic and not use the term race as a subgroup of *Homo sapiens.*

Chapter 3
Species

"Is there an official list of species? Yes, many of them —and they all differ from each other."

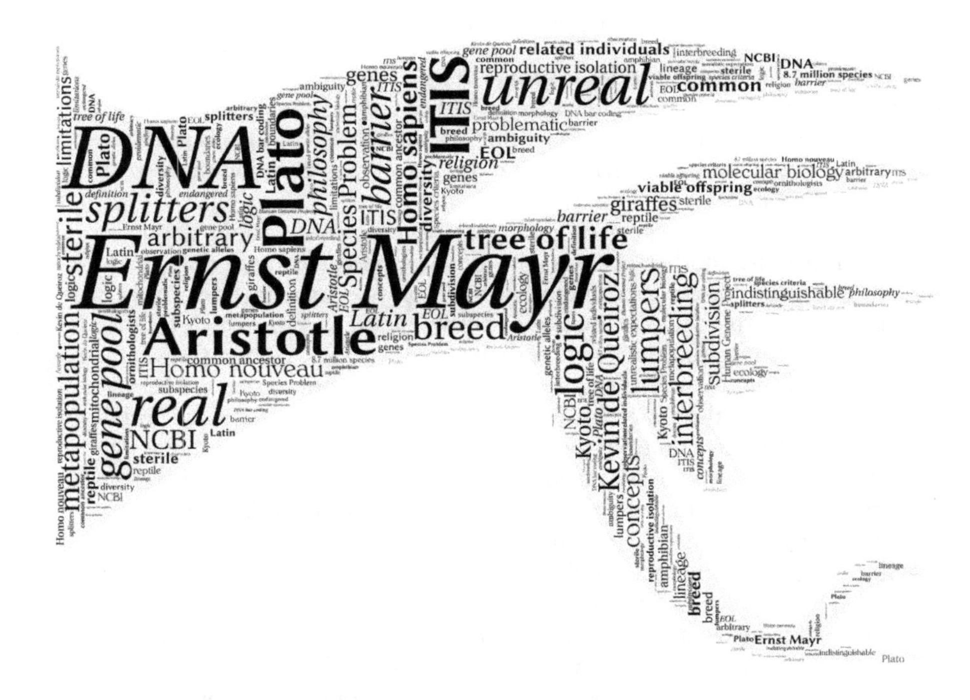

What exactly is a species? When I started the research for this book, I had not anticipated how much time and effort I would have to devote to this question. Probably like you, I just assumed there was a widely accepted answer. There isn't. It is a question that has been debated at least since the time of Plato, Aristotle, and other ancient Greeks.[10] The debate through the past two millennia has been intertwined with philosophy, religion, and logic, and more recently with genetics and molecular biology. If you expect a definitive answer after my research, you will be disappointed. Nonetheless, if we are going to talk about what species comes after *Homo sapiens,* we need to agree on what we mean by the word species.

The species problem

A typical definition of species is provided by dictionary.com: "the major subdivision of a genus or subgenus, regarded as the basic category of biological classification, composed of related individuals that resemble one another, are able to breed among themselves, but are not able to breed with members of another species."

Really? The "basic category of biological classification" sounds good, but the "resemble one another" part is a bit iffy. I guess you can say that all *Homo sapiens* resemble one another in that we walk on two legs, have relatively little hair, have opposable thumbs but not opposable toes, and have high foreheads (see figure 2). But so did the Neanderthals. Look at dogs. Does a Great Dane really resemble a Chihuahua (figure 3)? They are the same species. There are many other species that resemble each other much more than people and dogs, but they are split into multiple species. For example, the Alaskan malamute and the grey wolf look pretty much alike but are two different species (figure 4). Any birder knows that there are pairs of species of birds that are nearly indistinguishable when held in your hand, yet they are classified as separate species. Take for example the Willow Flycatcher and the Alder Flycatcher in figure 5. They are two different species distinguished primarily by their vocal calls (and some genes that we can't see). But different people have different languages (and some genes that we can't see), and we don't use that to separate them into different species. Then there is the Eurasian golden jackal, which lives in Europe, Asia, and Africa (figure 6, left picture). Recently, taxonomists decided that the one that lives in Africa (figure 6, right picture) is a separate species. Finally, to really throw a wrench in the "resemble" idea, figure 7 shows two pictures of the same species at different times in its life cycle: the monarch butterfly in its butterfly stage and its caterpillar stage. (You'll have to wait until chapter 5 to understand this one.) This "resemble" business seems awfully arbitrary, inconsistent, and unscientific.

Figure 2

Same Species - *Homo Sapiens*

Figure 3: Same Species

Chihuahua and Great Dane—*Canis familiaris*

Figure 4: Different Species

Alaskan Malamute—
Canis familiaris

Grey Wolf—
Canis lupus

Figure 5: Different Species

Willow Flycatcher
Epidomax traillii

Alder Flycatcher
Epidomax alnorum

Figure 6: Same Species Until They Decided They Were Different Species

Eurasian Golden Jackal
Canis aureus

African Golden Jackal
Canis anthus

Figure 7: Same Species

Monarch Butterfly
Danaus plexippus

Monarch Butterfly Caterpillar
Danaus plexippus

And how about that inability to breed between species? Wolves and dogs do it all the time. Nine percent of all bird species interbreed with other species.[11] Interbreeding between species is actually fairly common. Somehow we need to clarify this definition.

That is not a simple task. Many books have been written about the definition of species, and debate still continues about its essence.[12] First, disavow yourself of the notion that a species is something concrete, clearly distinct from other species, and stable. It is none of the above. Instead, a single species has significant variation within its population as we see with dogs and humans. Some may have fuzzy and often debatable boundaries from similar species. Any birder knows that at any given time, there may be conflicting definitions of the species, and species certainly change over time. Changes over time may be real biological and evolutionary changes, or changes in interpretation and classification by members of the one species capable of doing so, *Homo sapiens*. This difficulty in defining species is appropriately referred to in the literature as the species problem.[13]

Is there a solution to the species problem?

In an opinion piece in *Trends in Ecology and Evolution*, the authors stated, "There are almost as many concepts of species as there are biologists prepared to discuss them." They even considered the possibility of eliminating the entire concept of species altogether. The alternatives, however, seem equally problematic. They concluded, "For the moment, at least, users must acknowledge the limitations of taxonomic species and avoid unrealistic expectations of species lists."[14]

With the advent of rapid DNA sequencing techniques, the study of species has evolved from the days of Darwin's finches. In the past, one could observe variations in organism morphology in both live organisms and fossils, study geographic and ecologic location diversity, observe interbreeding habits, and determine a species. That's how the giraffes in Africa were all lumped into a single species with nine subspecies distributed geographically throughout Africa. Today, one must also perform complex statistical analyses of DNA sequences. When they did that with the giraffes, they concluded (to their surprise) that there were really four distinct species of giraffes and at least two of them should be put on the endangered species list since there were so few of them.[15]

Has the species problem been solved by DNA analysis? No. With this advancement in the tools created by *Homo sapiens*, the species problem has, if anything, become more problematic. The Human Genome Project[16] has identified over twenty thousand genes in *Homo sapiens,* with billions of variations of these genes within our one species. Which of these variations will be used to help define *Homo nouveau* and how that will be done are yet to be determined. Are there already multiple species currently classified as *Homo sapiens*? We'll come back to this in later chapters.

One of the more widely quoted definitions of species referred to for decades in textbooks and the literature was provided by Ernst Mayr more than a half century ago: "Species are groups of actually or potentially interbreeding natural populations, which are reproductively isolated from other such groups."[17] This definition focuses on "natural" biologic tendencies of populations in their ecological settings, and it stresses reproductive isolation. Although this was the most widely accepted definition for decades, it leaves room for interpretation and debate. What does "potentially interbreeding" mean? What is a "natural population?" What is reproductive isolation? Indeed, by the end of the twentieth century, dozens of other definitions had been published focusing variously on morphologic, genetic, and ecological factors, among other criteria.[18] Appendix 1 shows one table that attempts to list all of the various definitions.[19] To some extent, these different definitions reflect the professional field of study or the particular subset of the domain under observation; they represent those canyons I mentioned earlier. For example, a paleontologist would rely heavily on bone shape, and a geneticist would rely heavily on DNA analyses.

More recent species concepts focus on a technique called DNA bar coding, which utilizes rapid DNA sequencing to identify and distinguish species. Usually this sequencing is done on a small segment of DNA that is a marker for species variation. One of the most widely used markers is the mitochondrial DNA gene for one particular enzyme.[20 21] Even more recently, we can now easily and cheaply sequence the entire mitochondrial and nuclear DNA for use in determining species as was done for the giraffes.[22 23] I'll describe these various types of DNA in chapter 5, but I can tell you now that they don't solve the species problem.

Species lists are arbitrary!

Who decides which definition of species to use? Is there an official list of species? Yes, many of them—and they all differ from each other. (See box for some examples.) Then there are numerous specialized lists of species. (Also

Examples of Species Lists

- ITIS: Integrated Taxonomic Information System—a US government organization that maintains an "official" online file of taxonomic information on plants, animals, fungi, and microbes.[24]
- NCBI: National Center for Biotechnology Information—a part of the National Library of Medicine that maintains an online taxonomy browser for genetic information on species.[25]
- Kyoto Encyclopedia of Genes and Genomes—a database of genetic information on species.[26]
- EOL: The Encyclopedia of Life project—a collaboration of multiple organizations to provide information and pictures of all species known to science.[27]
- Species 2000—a federation of database organization that includes ITIS, NCBI, and EOL.[28]
- American Ornithologists Union's Committee on Classification and Nomenclature of North and Middle American Birds.[29]
- Amphibian Species of the World maintained by the American Museum of Natural History.[30]
- Reptile Database.[31]

in box.) All of these various organizations rely on committees of experts to review the taxonomies in their specialties and make decisions regarding additions, deletions, species splitting, and lumping. There is no one authoritative list of species. This leads to confusion in the scientific literature as well as public policy related to protecting endangered species. Do you want a list of species? Pick one.

Researchers at the University of Michigan and ten other institutions have put together an online graphical database they are calling the tree of life.[32] It is a taxonomic representation of all species known to be living today, with references and other information that can be browsed online. It contains 2.3 million entrees obtained from many sources. The intent is to continually update it and keep it available online. That will be quite a big job because it is estimated that there are a total of 8.7 million species alive today, which means that 74 percent of them are still unknown.[33] The ability to graphically browse through the various taxonomies in this tree of life is a valuable resource, but there is nothing official about it and it certainly doesn't solve the species problem.

Are you a lumper or a splitter?

When it comes to species (as with any classification system), there is a tension between lumpers and splitters. For example, at one time, taxonomists defined a warbler species called the Audubon's warbler and another species called the myrtle warbler. These two species had a different, but overlapping geographic range. They also differed in color, in that the Audubon's warbler has a yellow throat, and the myrtle warbler has a white throat. After examination of their interbreeding patterns in various geographies, in 1973 the American Ornithologists Union's committee concluded they were really a single species and lumped them together as the yellow-rumped warbler, demoting each of the original species into subspecies. However, in 2016, after a more complete genetic analysis, researchers are now recommending that they again be split into two separate species.[34] Are these species changing or is it simply *Homo sapiens'* tools that are changing? Or maybe it's just the committee members that are changing.

This brings up the concept of subspecies. Although a legitimate taxonomic term, the definition of a subspecies is even more vague and difficult to define than the definition of species. Generally, subspecies can interbreed with each other and produce viable offspring, but tend to exist in separated localities and therefore do not have much opportunity to interbreed. They are clearly constructs of taxonomy committees. In the taxonomy of naming, one adds a third Latin term (or let me say, a term that is made up to look like Latin) to indicate the subspecies. Thus the old Audubon's warbler is *Setophaga coronata auduboni,* and the old myrtle warbler is *Setophaga coronata coronata.*

Are there any subspecies of *Homo sapiens*? Yes, but only one exists today, *Homo sapiens sapiens*, and we are it. As far as I can tell, this subspecies designation is used rarely in order to distinguish it from the extinct *Homo sapiens idaltu*, which some believe immediately preceded us in evolution. We will revisit this question in the discussion of human races in the next chapter.

In general, the splitting of species rather than lumping is the more common issue that creates debate and complexity. The splitting of species is how the evolutionary tree grows. It is why we have a common ancestor with the apes, and why every living thing ultimately has a common ancestor. It is why we have millions of different species today, all of which differ from the billion or so that have gone extinct.

But depending on how you define a species, at any given time with any given species, the splitters could derive a variable number of new species. An illustrative example is described in a paper on a species of California trap-door spiders. Depending on which species criteria are used, this species could be split into anywhere from five to over sixty different species without any actual evolution taking place![35]

If you really want to get confused about the whole notion of species, read John S. Wilkins book *Species—A History of the Idea*[36] or David N. Stamos's book *The Species Problem*.[37] The more I read, the more confused I got.

Are species real?

There is a debate as to whether or not species are real or simply concepts that exist only in the brain of *Homo sapiens*; a longer discussion of this debate is contained in appendix 2. My own belief is that species are concepts created by humans rather than entities that have some discoverable existent realty. For most species distinctions, there is little ambiguity. No one argues that a snake and tiger are the same species. It is when species are similar in appearance that fine, and somewhat arbitrary, distinctions must be made in classification. That is not to say that there is no objective observation of the organisms and their fossils in nature that informs our decisions regarding species. There certainly is. Taxonomy is a science of careful observation and study.

Whether species are real or concepts, there needs to be agreement on how we will define the future *Homo nouveau*. I think John Brookfield said it best.

> *The essence of the 'species problem' is the fact that, while many different authorities have very different ideas of what species are, there is no set of experiments or observations that can be imagined that can resolve which of these views is the right one. This being so, the 'species problem' is not a scientific problem at all, merely one about choosing and consistently applying a convention about how we use a word. So we should settle on our favorite definition, use it, and get on with the science.*[38]

So I will get on with it. Whether species have existent reality or are concepts in our minds, everyone agrees that there is a species called *Homo sapiens*, and one day another species (which I'm calling *Homo nouveau*) may emerge from it. I'm looking for a definition of species that fits the ability to define and distinguish these two species. The danger is that because the very definition of species is subjective, this search for a definition could be biased. Specifically, I could choose a definition that makes it either easier or more difficult to establish *Homo nouveau* and distinguish it from *Homo sapiens*. I will attempt to select a definition based on relevant and established criteria in the scientific literature. But once I do, it will be a judgment call and subject to debate.

The (arbitrary) definition of species that I'll use in this book.

Kevin de Queiroz has created a diagram (figure 8) that illustrates the problem in defining a new species and also suggests a way to resolve the problem.[39]

FIGURE 8

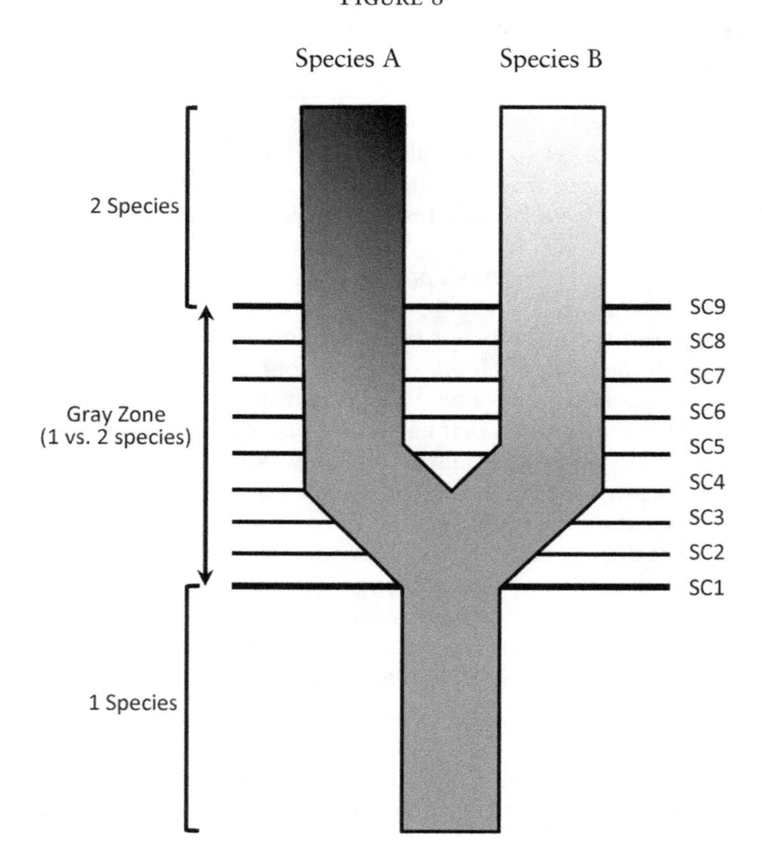

From: Kevin de Queiroz, "Species Concepts and Species Delimitation," *Systematic Biology* 56, no. 6 (2007): 879–886.

The black area is Species A, and the white area is Species B. The gray area represents uncertainties in the literature regarding the existence of Species B. SC 1 through 9 represent different criteria (SC stands for "species criterion") for defining a species. They could be criteria such as morphology, reproductive isolation, or specific genetic alleles. Kevin de Queiroz proposes in this paper that one can eliminate this gray area (and thus the species problem) by going to a more fundamental or basic definition of species and using the factors described in SC 1 through 9 (or whatever number) not as necessary or sufficient criteria, but rather as further possible descriptors to more clearly define the species if and when the species criteria become known. The fundamental or basic definition he proposes is that *a species is a segment of a separately evolving metapopulation lineage.*

That sounds complicated (and it is), so I will try to explain it. A metapopulation is a subset of the total population of organisms sharing a common gene pool; that is, they share genetic characteristics that enable them to interbreed with each other in a manner that produces viable offspring of either sex. In other words, they can have children who in turn can have more children. The lineage component of the definition implies that the species is not simply a snapshot at any given time; it occurs over time with ancestors and descendants of the same species that are "separately evolving."

It is the "separately evolving" part of this definition that is the most vague in my view. How will we know that? We will need to demonstrate that the population has changed (or not changed) over time in the same or very similar manner. That is, whatever the SCs are that characterize a species, they change (or not change) in the same way over time. Although I struggled with this problem, you will see in later chapters that it can be resolved.

The metapopulation does not need to be in a single location. It could consist of multiple pockets of individuals living in different locations, as long as the different pockets evolve in the same way as each other but differently from other metapopulations. Any particular SC may or may not apply to the species so defined. But when it does apply—for example, a certain skull configuration, brain size, or particular DNA sequences—that SC becomes a part of the species definition going forward, but it was not necessary to define the species in the first place. Members of the species may or may not interbreed with some members of other species. That is, reproductive isolation could become an SC of the species (and usually does), but that is not necessary to define the species initially. What is necessary is that breeding within the species produces viable offspring, and that they evolve separately from other species.

This definition of species purports to resolve the ongoing conflicts between advocates of different concepts of species. It takes these different concepts and turns their varying criteria (morphology, genetics, ecology, etc.) into secondary characteristics the list of which could change over time to further define a species and delimit it from other species, but it does not require any of them to define the species initially. These secondary characteristics may also lead to the identification of subspecies, or eventually to the splitting of the species into two species. I frankly don't think this really solves the species problem (because it is not solvable), but it is the best definition I could find.

The result of the de Queiroz definition is a lower threshold for declaring new species at an earlier point in time, and it also creates species that are less clearly delineated from other species. I will come back to these points later when discussing *Homo nouveau*.

Chapter 4

Speciation

"We aren't very big, don't run very fast, can't fly, can't swim very well, and aren't very strong compared to lots of predators out there. Our brain functions were selected to overcome all of that and allow us to live long enough to make babies."

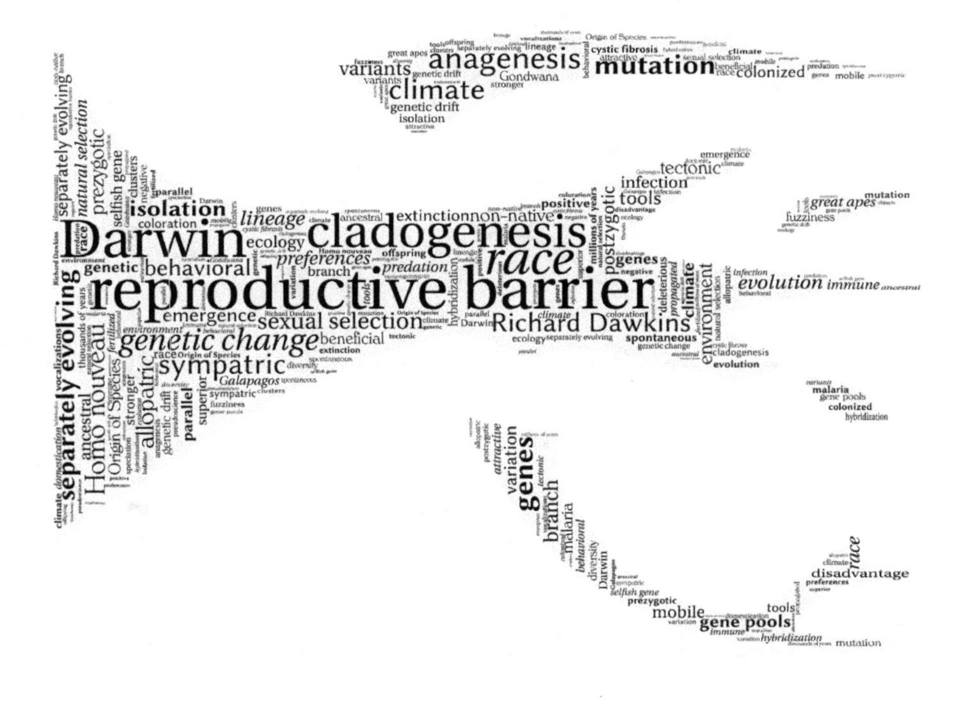

If and when *Homo nouveau* appears on earth, it will do so through a process called speciation. Speciation is the formation of new and distinct species in the course of evolution.[40]

As will be discussed in chapter 12, the concept of "the course of evolution" may change dramatically when it comes to the emergence of *Homo nouveau.*

Historically, speciation has occurred over long periods of time, usually thousands to millions of years. In the vast majority of instances, the new species appears as a branch from an existing (ancestral) species (figure 9), where the ancestral species continues to exist in parallel with the new species at least for some period of time. Either the ancestral or the new species may die out at some later time or spawn other new species. This process is called cladogenesis.

FIGURE 9

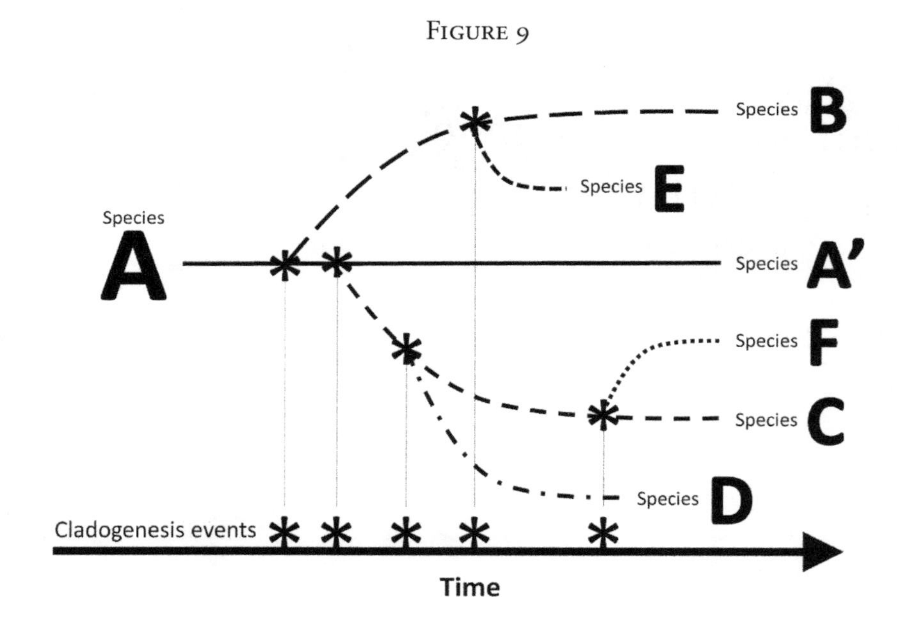

Cladogenesis, Anagenesis and Hybridization—Some New Words

It is possible that a single species could evolve such that there are major differences over time between that species at a given time compared to some previous time; this is called anagenesis. Species A' in figure 9 is the same species as Species A in the figure, but it differs because of anagenesis. In this case, Species A' did not branch off in parallel to the ancestral Species A. The differences between this single species at a later date compared to how it was at an earlier date can be as great or greater as those between two different species. We are dealing with different points in time, so there is no way to know if interbreeding would be possible between the two versions of this same species or what their offspring would be like. Thus we typically do not label the later version as a new species; rather, we consider it the evolutionary result of the same earlier species.[41] It is quite possible that the *Homo sapiens* of today could not interbreed with the *Homo sapiens* of two hundred thousand years ago, but we have no way of knowing this. When looking at the fossil record, which is usually sparse and has large time and geographic gaps, it is not always clear whether to interpret the changes in a later fossil as a new species by cladogenesis, or simply the evolution of the same species by anagenesis. This is particularly true in our lineage.

Another mechanism of speciation is the combination of two ancestral species into a single species; this process is called hybridization and is rare in animals.[42] The terminology here is somewhat confusing in that hybrids, which are the offspring of the mating of two different species, is a fairly common event, however the creation of new species by hybridization is a rare event. For example, when wolves and dogs (two different species) sometimes interbreed, the offspring of such matings are not a separate species from either wolves or dogs because they do not continue to evolve independently. So what species are these offspring? Technically, they are simply hybrids and belong to no species, or they can be considered variants of either wolves or dogs.

An example of true hybridization was reported for two species of a genus of fruit flies that occurred after a nonnative plant was introduced into their environment. The two species colonized the new plant and ultimately hybridized into a new species evolving independently from its two predecessors.[43]

The lumping of the myrtle warbler and the Audubon's warbler into the yellow-rumped warbler, discussed in the previous chapter, is probably not an example of a new species being created by hybridization, but rather simply a change in classification based on further study by taxonomists. It is a matter of the splitters winning the battle years ago when they declared the Audubon warbler and myrtle warbler were two species, and then the lumpers won the battle in 1973. It looks like the splitters may reverse this again, but I would not call that speciation by cladogenesis. It is simply *Homo sapiens* re-thinking the issue. In any case, because I'm considering here only the possible emergence of *Homo nouveau* from *Homo sapiens*, hybridization is not relevant to the discussion. There is no other species with which *Homo sapiens* can realistically hybridize, at least as *Homo sapiens* is currently defined by all of the many definitions.

In summary, when I consider the possible speciation of *Homo nouveau* from *Homo sapiens*, I focus on a possible cladogenesis event rather than anagenesis or hybridization.

Is speciation real?

Whereas species may only be a concept, speciation is real. That seems initially like a logical absurdity. How can speciation be real if species isn't? I have no problem with it. No matter how one defines species, how many books have been written about it, and how many different opinions there are on its definition, there is no doubt that new species have emerged over the course of time. That is, new species (however defined) have emerged not only in our thinking but also in reality. To state it differently, organisms are clearly real. They evolve and change in reality to an extent that no matter how one classifies these organisms into species, new species emerge. There is no debate that speciation occurs. For example, fifteen million years ago, no great apes existed. They do now. Obviously all of the great ape species emerged at some point in time. I will focus our attention on the possibility that *Homo nouveau* will really emerge from *Homo sapiens* by some form of cladogenesis speciation.

Genes and barriers

Ultimately, speciation must involve genes. Our working de Queiroz definition requires gene pools to be separately evolving. A new species must, by definition, have something different at the genetic level from the ancestor species from which it emerged. But what must that genetic difference be? We know that there is great genetic variation within a single species. Simply having a genetic difference, although necessary, cannot be sufficient to define a new species.

It is the "separately evolving" piece of the definition that must also be met. That is, there must be something about the genetic difference or the environment that causes those members with that new gene pool to evolve independently of the species from which it emerged. If large portions of the members of an emerging gene pool interbreed with the ancestral members to a sufficient degree, the gene pools would become so mixed that they would no longer be evolving separately. That is what is happening today among the vast variety of *Homo sapiens*. Therefore, another way to think about speciation is to consider what barriers might be present or become present that would prevent or greatly reduce the amount of interbreeding between members of the emerging gene pool and their ancestors. In a sense, then, "the *process* of speciation involves acquiring reproductive barriers."[44]

Two things must happen for speciation: a change in genes, and formation of barriers to interbreeding between two populations of the species. These do not have to happen simultaneously or in any specific order. That is, the barriers can come either before or after the genetic changes. They also do not have to be complete barriers, as we have seen already with some interbreeding species.

Genetic changes are occurring constantly in every species, including our own; we'll understand how those occur in the next chapter. Most of these changes do not result in a new species. Instead, they result in a tremendous amount of variation within a single species. How, then, do we get to the point where there is a population with a gene pool of these changes that are evolving separately from all other species?

Natural selection

This next discussion usually begins with reference to Darwin's *On the Origin of Species*.[45] Darwin introduced the concept of natural selection as the primary mechanism driving evolution. Darwin accepted the fact that organisms change in some spontaneous manner, although at that time DNA, genes, and the processes that will be discussed in the next chapter were not known. His premise was that nature would "select" from among competing organisms those that were best adapted to survive in their environment and reproduce. That is, if the change that occurred in an organism made that organism more likely to thrive—specifically more likely to reproduce—whatever that change was would be more likely to be propagated into future generations. Put into today's terms, natural selection favors the propagation of genes that imbue some reproductive benefit. It is not the genetic change that imbues that benefit, but the genetic change interacting with the environment.*

For example, if a mutation changed a gene that made an organism less likely to get malaria, that mutation would only be selected for propagation if the organism lived in an area where malaria was commonly present. Otherwise, that gene would not be selected. And even if the organism did live in an area with malaria, it would not necessarily lead to a new species as long as the individuals with that genetic trait continued to interbreed freely with those without it. In fact, we have that exact example in *Homo sapiens*. Humans with blood type O are less likely to die from a common form of malaria than are people with blood types A or B.[46]

So why doesn't everyone who lives in a malaria region have blood type O? There are several possible reasons. Perhaps the genetic change causing the different blood types hasn't been around long enough for this selection to become apparent. Or perhaps the other blood types convey other advantages that lead to their selection. Or perhaps blood type O also conveys some disadvantage regarding other diseases that counteracts the beneficial effect. Nothing is simple.

There are many forces in nature that can act as natural selection forces and many ways organisms have adapted. (See box). The list is endless. The random genetic changes that led to our superior brain were selected because a superior brain allows us to make tools, control fire, communicate with language, kill at a distance (with spears, arrows, bullets, etc.) We aren't very big, don't run very

*Darwin's theory is often characterized by the phrase "survival of the fittest." I am not a fan of that characterization because the definition of fitness simply comes down to those traits that survive—a tautological concept.

Examples of natural selection forces.	**Examples of natural selection adaptations.**
◗ Infectious agents	◗ Resistance to infection
◗ Predators	◗ Speed
◗ Competitors in food chain	◗ Strength
◗ Climate	◗ Sensory ability
◗ Food supply	◗ Being poisonous
◗ Water supply	◗ Body armor and shells
◗ Homo sapiens	◗ Intelligence
	◗ Metabolic adaptations

fast, can't fly, can't swim very well, and aren't very strong compared to lots of predators out there. Our brain functions were selected to overcome all of that and allow us to live long enough to make babies.

Richard Dawkins, author of the book *The Selfish Gene*[47] would argue that my language describing natural selection is imprecise. When I say "our brain functions were selected" or "organisms are selected" that I am implying that natural selection acts on the level of a whole organism or even an organ within an organism. That is incorrect. He states that natural selection acts on genes and only genes. It takes thousands of genes to make our brain function and many more to make our whole body function. To Dawkins, our body is simply a vehicle to convey a gene to the next generation. Although many genes need to cooperate to create an organ or whole body, it is the variation in single genes that ultimately gets selected or not by interacting with the environment. Natural selection acts at the gene level—not the organ or whole organism level and certainly not at the level of groups of organisms or whole species.

Virtually everything in the ecology in which an organism exists in some way creates a natural selection force. Again, it is the random genetic change that leads to new features and functions of an organism *combined with* something in the environment that leads to either a better or worse ability to reproduce; that is the essence of natural selection.

Even *Homo sapiens* itself is a natural selection force. When we domesticated wolves, we created dogs. Sometimes human domestication of plants and animals is called artificial selection. Nonetheless, this selection force acts the same as any other national selection force. We move plants and animals to new locations. We change ecologies in ways that lead to the extinction of many species. We have become a major natural selection force, both to create new species as well as to cause the extinction of existing ones.[48]

Changes occur over time in the genome of an animal bred for any particular trait. Sometimes the same breeding produces other traits that were not intended, indicating that the selection force acts more broadly than on any single gene or even sets of related genes. For example, foxes bred for tameness also develop changes in the color of their fur.[49] Selective breeding is an imprecise tool.

Sometimes Environment Change Does not Lead to Adaptation.

Today's natural selection forces are dramatically different from two million years ago, when the *Homo* genus emerged, or even two hundred thousand years ago when *Homo sapiens* began to flourish. For example, two hundred thousand years ago, when we were still hunters and gatherers roaming among many stronger and more capable predators, our brains were tuned to react quickly to the slightest movements and possible threats around us. Our environment has changed since then, yet this phenotypic behavior may not have evolved to accommodate the fact that we don't require this level of hypersensitivity to our surroundings. What was once a survival mechanism is today a distraction—or worse, the cause of attention deficit/hyperactivity disorder. The random mutations needed to make us less sensitive have not happened. Similarly, our excess storage of fat was important back then as a backup for periods of food scarcity. Today, that capability is no longer needed and is a cause of obesity. Our environmental changes have outpaced our evolutionary adaptation process in some ways.

Natural selection works both positively and negatively.

That is, some random genetic changes, when interacting with the environment, increase the possibility of survival and reproduction. This is positive selection.[50] Most random genetic changes do not have any effect because they don't occur in an area of the genome that is important, or they don't change the function of a gene. However, some make an organism less viable. These probably outnumber the positive changes, but because the offspring tend to die before reproduction age, these genetic changes are not propagated to future generations and tend to disappear from the gene pool. This is called negative selection. The question arises, then: Why do we have any surviving genetic diseases? Why doesn't negative selection eliminate all of them?

Why doesn't natural selection get rid of genetic diseases?

Negative selection (and positive selection) works over many generations. It takes time, sometimes millennia or longer for negative selection to weed out the bad genes that arise randomly. Since mutations are occurring constantly, it is no surprise that at any given time in the evolution of any species, there are lots of bad genes around. Nonetheless, some genetic diseases persist over time without decreasing in numbers. Why?

The answer is complicated and not known in every case. One explanation is that the negative gene does not become active until after the reproductive age of an organism. In humans, that is the case for a genetic disorder called Huntington's disease, which only becomes manifest in older age. By the time Huntington's disease becomes symptomatic and lethal in patients, they have already passed on their destructive Huntington's disease gene to some of their children.

Another explanation is the case of cystic fibrosis. This is one of the most common genetic disorders, with approximately 4 percent of the population being carriers of the genetically abnormal gene. A carrier (meaning only one of an individual's pair of genes is abnormal) does not have the disease. You must have two copies of the abnormal gene to have the disease. (We will discuss genetics in more detail in the next chapter.) Nonetheless, because cystic fibrosis often leads to death prior to reproductive age, one would have expected negative selection to remove the gene from the human gene pool over the millennia.

It turns out that carriers of the cystic fibrosis gene may be more resistant to tuberculosis and other infectious diseases, which would generate an offsetting positive selection for the gene.[51] Similarly, people who have sickle cell trait (i.e., contain only one copy of the defective gene) rather than sickle cell disease (i.e., contain two copies of the defective gene) are more resistant to malaria, which partially offsets the negative selection for this gene.

Finally, there is a statistical process, called genetic drift, that could account for the persistence of some genetic disorders. Genes and gene combinations change over time in a random fashion (described in the next chapter). As stated above, some of these changes confer survival benefits to an organism; some are neutral, and some are deleterious. Statistically, a proportion of beneficial changes will not be propagated into future generations by chance, whereas some deleterious changes will be, in spite of natural selection. Given the large number of deleterious changes that randomly occur, statistically a small portion of them will penetrate a population in large numbers.

There is no goal to natural selection.

The use of the word "goal" implies a process with an end result in mind, and that is misleading. It isn't as though someone or something a billion years ago said, "Wouldn't it be nice if an organism could hear by detecting sound waves?" and that somehow led to the development of the ear. It didn't work like that. Sound detection in humans utilizes three bones in our middle ear to transmit the vibrations from our eardrum to our inner ear. Those three bones evolved from existing jawbones in reptiles and fish that had nothing to do with hearing. They involved random mutations. The process is call exaptation. Similarly, the rods and cones in our eyes that detect electromagnetic radiation for sight evolved from other more primitive nerve endings.

Although random mutations have led to some *de novo* innovations in evolution, most innovations in body design have come from mutations causing exaptation of existing structures or functions. That is true of feathers used in flight, limbs used in the first land dwellers, and the advanced cognitive abilities of our brains.[52] The human pharynx (voice box), thyroid gland, jaw, and other structures evolved from ancient gills. The random mutations that led to the exaptation of jawbones into middle ear ossicles happened without any relation to hearing; they just happened. The fact that these bones responded to sound vibrations was a random occurrence. But once that happened and the organism gained some reproductive advantage by being able to better detect sound waves, this feature was naturally selected to be passed on to future generations, and it allowed other random mutations that improved the process to further enhance hearing.

One could logically state that it would take more than just some bones to vibrate in response to sound waves for hearing to occur. There would need to also be some nerve connection to some part of the brain that could detect that these bones were vibrating. That is true. But it is likely that some primitive nervous system already was in place that detected sound vibrations that impinged somewhere on the organism, and the bone ossicles simply improved that detection.

Natural selection is not the whole story.

Although natural selection is a major process involved in speciation, it is incomplete as an explanation of speciation. It is more an explanation of evolution of a species rather than speciation. If the members of the population containing the naturally selected gene continue to interbreed with those without the gene, the gene pool will not evolve separately. The species will evolve but will not branch into a new species. For example, people who are carriers of cystic fibrosis are not a separate species; they interbreed freely with all other people. Something else must occur to isolate the evolving species from its ancestors so that it evolves separately. That something else is the barrier to reproducing with other species, i.e., reproductive isolation. Although natural selection is one force that leads to reproductive isolation, it is not the only one.

Barriers to interbreeding

To the layperson, the literature on reproductive isolation barriers is as complicated as the literature on species definitions. Appendix 3 contains a list of possible reproductive isolation barriers. Many of the categories in appendix 3 would not likely apply to the emergence of *Homo nouveau,* so I will focus on a smaller set of speciation factors that are relevant to the answers discussed in chapter 14.

Two broad categories of speciation are called allopatric and sympatric. Allopatric speciation occurs when species are physically or geographically separated. This separation can be huge, such as when the movement of earth's tectonic plates separated Australia from the rest of ancient Gondwana. They can be smaller, as in the separation of the various Galapagos Islands. Or they can be even smaller, as in the separation of two adjacent valleys or across a river. They can be even smaller yet, as in the separation of treetops from tree roots. The point is that there is some physical barrier that has been introduced into a single species that divides it into two populations that cannot physically interact with each other.

That physical barrier can come to be in many ways, and some were already mentioned above. One common way is for some subset of the species population to migrate to a different area; perhaps in search of a better food or water supply, perhaps due to changing climatic conditions, perhaps because it was forced away by a competitive species—there are many possible reasons. There are examples of all of these. The separation could occur by accident—for example, being blown to a distant location in a storm, being swept to sea, or riding in the fur of a migrating animal. Finally, humans can and do often introduce subpopulations of existing species into geographically distant locations.

Separating a portion of a population into a new location does not create a new species. At least at first, they still are a part of the same gene pool. They could even continue to evolve over time in the same or similar manner such that they continue to meet whatever definition of species is being applied, and they do not lose the ability to interbreed when members from each location are brought together. For example, red deer exist today on multiple continents. Some populations are clearly physically separated from others yet remain part of the same species. They together represent a single, separately evolving gene pool. We can find similar examples among *Homo sapiens.* The aborigines that populated Australia were isolated from other *Homo sapiens* for thousands of years, yet when the explorers from Europe arrived, they were able to interbreed without difficulty and produce viable offspring.[53] In fact, all of the various forms of *Homo sapiens* today spread out over many geographic areas and ecologies represent one combined, separately evolving gene pool. During my research, it made me wonder whether *Homo nouveau* could ever emerge!

There are several ways that allopathic populations can diverge into two species. Probably the most important role is natural selection.[54] Because the two different locations may have different ecologies, there could be divergent natural selection forces. Even if the ecologies of the two locations are the same, different mutations in the two populations could make them evolve differently in response to the same natural selection forces. Over time, the random changes to the genes of each population can diverge in a manner that eventually creates some barrier other than geography to prevent the two populations from successfully interbreeding. This divergence of two gene pools happens more rapidly when the populations are separated than when they are not separated, because there is no opportunity for the two gene pools to mix.[55] Note that natural selection did not necessarily play any role in the two populations becoming separated in the first place, and it may not play any role in their divergent genetic changes leading to reproductive isolation. On the other hand, natural selection could and usually does play a role at either time.

Sympatric speciation is more relevant to us.

The other major category of speciation is sympatric speciation. Sympatric speciation is when a species emerges within the same geographic and ecological area as the ancestral species. There is no physical barrier. It is intuitively more difficult to understand than allopatric speciation.

However, sympatric speciation is the likely situation that will apply to the emergence of *Homo nouveau.*

Most of the categories listed in appendix 3 can be causes of sympatric reproductive isolation. That is, there are many factors that may lead to the emergence of a new species from an existing species without being physically separated. It is a long and complicated list. I will focus on the few that are relevant to the emergence of *Homo nouveau*.

Sexual selection

One of the more interesting and potentially relevant factors is called behavioral isolation or selection. Simply put, this is when members of one sex of a species choose not to mate with members of the other sex because of some physical or behavioral trait. These sexual preferences, also called sexual selection, are common in many species where, for example, the male has developed certain coloration or vocalizations, or other physical attributes that make him more attractive to a female than other males. This is common in birds, butterflies, and many other species (including, to some extent, our own).

These sexually attractive or unattractive features can be very subtle or very flagrant, at least to the human eye. If anyone has watched the mating ritual of the sage grouse, one can see the elaborate lengths to which evolution has gone that allows a male sage grouse to strut, display, and pump up his air sacks in order to show the female how beautiful he is. Although this is all very visible to the human eye, we still cannot tell exactly why one particular male gets chosen over any others; only a female sage grouse can do that.

Whatever these sexually attractive features may be, they are genetically determined and can change over time. If the changes make them more attractive, then over time sexual selection can occur to create a new gene pool that evolves separately from its ancestor. The barrier is attractiveness (or lack thereof) to one subset of a female population that prefers males with the new physical feature, compared to the original population of females who still prefer males with the old feature.

One puzzling aspect of behavioral isolation is that it seems to require near simultaneous genetic changes in both the male to develop the new physical trait, and in the female to develop a preference for the new trait.[56] We'll come back to behavioral isolation later regarding speciation within the genus *Homo*.

Postzygotic isolation

Another set of speciation mechanisms falls under the category of what is called postzygotic isolating barriers (see appendix 3). A zygote is a fertilized egg. In humans, a zygote is formed when a sperm penetrates an egg. Postzygotic refers to any time after mating has occurred and a fertilized egg has been created. This includes the entire time of the pregnancy but also any time after birth as well. (The behavioral isolation barrier discussed above would be considered *prezygotic* because this barrier prevents successful mating and the creation of a fertilized egg.) Normally, genetic changes are constantly occurring in a species. Normally when new members of the population are born with a specific genetic change or set of changes, these new members can freely interbreed with all other members of the species to produce viable offspring. These new genetic changes then become part of the mix of genes propagated by the original ancestral species.

But sometimes the new genetic change leads to a problem when an individual with that change interbreeds with individuals without the change. If the resulting offspring (called hybrids) are either sterile or physically or behaviorally less able to reproduce, then the new gene pool will die out. But as long as the individuals with the new genetic change can interbreed with each other and produce viable offspring, a new species can emerge that is reproductively isolated, even in a sympatric environment.

Barriers don't have to be complete.

Just as there remains fuzziness in the concept of species, there also remains fuzziness in the concept of speciation, particularly in the requirement that there be some barrier to reproductive interbreeding. The barriers discussed above are not necessarily absolute barriers; they often are relative barriers. For example, the postzygotic barrier in which hybrid offspring are sterile is not always complete. Sometimes the offspring can reproduce but in a diminished capacity. Thus there is a continuum in the degree of this barrier in hybrids from complete sterility to complete reproductive capability.[57]

Speciation doesn't occur in a single generation. A mother doesn't suddenly give birth to a child of a different species. As the famous paleontologist Donald Johanson said, "There is never a clean break in a line of evolutionary descent."[58] Things will go along somewhat ambiguously for many generations, maybe even thousands of years, and although at some point it will be clear that a new species has emerged, placing the exact point in time when that happened is arbitrary.

Does Homo nouveau already exist?

Looking at today's *Homo sapiens*, has speciation already occurred into *Homo nouveau*? Certainly there is great genetic diversity between certain subpopulations of people. Clusters of subpopulations exist in specific locations with similar genetic backgrounds that interbreed largely within their own subpopulation. There are sexual preferences throughout our species that could have led to prezygotic sympatric speciation. Certainly there are people living in somewhat geographically isolated regions of the world suggesting that the conditions might be ripe for allopatric speciation, or that may have already happened long ago.

This brings us to the question of races. Do the variants of today's humans we call races constitute separate species? Certainly, skin color or epicanthic folds are examples of genetically determine human traits that are associated with certain races; but there are other genetically determined human traits like eye color, lactose intolerance, sickle cell anemia and thousands of other traits that tend to run in races but are not used to categorize races. Are races biological concepts or social concepts? Can we separate the controversies, biases, and sensitivities associated with discussion of race from the sciences of genetics, species and speciation? Not completely. In an editorial in the prestigious *New England Journal of Medicine,* Dr. Robert Schwartz stated that "Race is a social construct, not a scientific classification" and that race is a "pseudoscience" that is "biologically meaningless."[59] On the other hand, Neil Risch and his colleagues argue just the opposite in a later issue of the same journal.[60] They review the significant legitimate genomic research that clearly shows how humans have genetically clustered into five continent-based groupings: African, Caucasian, Asian, Pacific Islander and Native American. They also point out that these genetic differences are biologically and medically meaningful. Are these different species? Subspecies? These genetic clusters are real, but whether they should be characterized as defining taxonomic entities or even "races" is highly debatable. Different genetic markers could be used to create different clusters of *Homo sapiens.* Certainly there are genetic variations within *Homo sapiens* and these genetic variations are certainly biologically meaningful.

I am not going to attempt to resolve the controversies regarding race in this book. The only relevant question for me is whether "races," however defined, relate to whether *Homo nouveau* already exists or not. As you will see in chapter 8 (and appendix IV-I), all of today's *Homo sapiens* around the world are descendant from a very small population that migrated from Africa between 50,000 and 120,000 years ago—perhaps less than 1000 individuals. Thus we started out with a very narrow genetic base. This small number of individuals spread out over vast distances and multiple continents in a relatively short period of time. Clearly, clusters of individuals were isolated from each other long enough for genetic variations to arise, mostly associated with the separate continents they inhabited. Our advanced genetic tools can quantify these variations today. However, *Homo sapiens* have remained mobile, in fact increasingly so, such that interbreeding has continually mixed these now large, continentally-based clusters. The offspring of these mixed matings are themselves fertile. There has been little time for speciation to occur. There is no reproductive barrier and no separately evolving human metapopulation today. We are all one species. *Homo nouveau* does not yet exist.

Is it possible that there is a small-population exception that has remained more isolated than these larger clusters? For example, there is a tribe of people in Peru called the Mascho Piro. They have been isolated from the rest of world for virtually all of their known history. They live in a remote isolated part of the Amazon jungle and have had very little contact with outsiders.[61] In fact, the Peruvian government has declared it illegal to have contact with them for fear that their immune systems will not be able to fend off infections from such contact. They certainly don't interbreed with anyone outside their tribe, and it seems that they could well fit the definition of a separately evolving metapopulation lineage.

How separate has not been tested genetically or in any other way. There are about fifty other such known tribes living in isolation in parts of South

America. Similar tribes that have emerged from isolation in the past have been shown to be able to interbreed with other populations successfully, so it is likely that they are still part of the total *Homo sapiens* gene pool. Keep in mind, *Homo sapiens* have only been in this part of the world about fifteen thousand years which is a miniscule period of time evolutionarily speaking.

In the relatively short period of time that *Homo sapiens* have existed, there has not yet developed that crucial barrier to reproduction that is a necessary ingredient to declare *Homo nouveau*. Interbreeding can and does occur between any of the subpopulations of humans, producing viable offspring. Thus our search for the answers is still valid within our current definitions of species and speciation. There is no known evolving metapopulation lineage of humans separate from the *Homo sapiens* population today. When there is, we will have *Homo nouveau*, and that is what I am attempting to envision.

Chapter 5

Genetics

"I told you this was far more complicated than Mendel's peas."

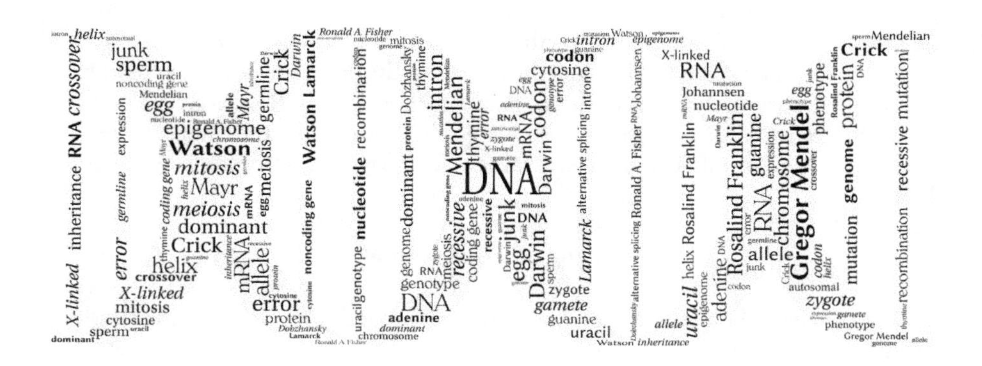

Darwin published his epic theory of evolution in 1859. He knew that organisms changed over the generations but didn't know how. Up to that time, it was thought that various traits were acquired during the lifetime of an organism, and that these acquired changes would be passed on to its progeny. It was thought that children were always a "blend" of their parents. The famous French biologist Jean-Baptiste Lamarck promoted this view. For example, he believed that giraffes evolved their long necks because previous generations of giraffes stretched their necks to reach higher leaves in trees, and over generations this stretching led to the inherited trait of long necks.

It wasn't until the mid-to-late 1800s that Lamarckian evolution was debunked and humans realized that there were discreet "units" of heredity that get passed on between generations. These units were not related to anything acquired by the individual after birth. This was based on Gregor Mendel's work with peas, where he disproved the notion that parental traits were blended in the offspring. Instead, these traits are carried either all or none according to specific rules of dominant and recessive inheritance. It wasn't until 1909 that these units of inheritance were named genes by Danish botanist Wilhelm Johannsen.

Gregor Mendel's work

Gregor Mendel observed that if he cross-pollinated a line of peas that had been consistently yellow over generations with a line of peas that had been consistently green over generations, all of their offspring would be yellow. But to his amazement, if he cross-pollinated strictly within the yellow offspring, one-quarter of the subsequent generation would be green, and three-quarters would be yellow. That is, the green color came back even though both of the parents were yellow. From this simple observation across many different types of traits, over many generations and many different plant species, he developed the concept of dominant and recessive inheritance traits. We call this Mendelian genetics or Mendelian inheritance.

Basically, Mendelian genetics states that we inherit two copies of every gene, one from the mother and one from the father. These genes can have either a dominant variant or a recessive variant. In the case of the pea color gene, the yellow variant is dominant and the green variant is recessive. If a progeny inherits the yellow variant from each parent, then the progeny will also be yellow. That progeny is called homozygous for the yellow trait because both of its genes for color are yellow. If the progeny inherits one yellow variant from one parent and one green variant from the other parent, the progeny will also be yellow because yellow is dominant. But this progeny will have one of each variant of the color gene, and so it is called heterozygous. Similarly, the people with sickle cell trait or the carriers of cystic fibrosis discussed earlier are heterozygous with regard to those disease genes.

Mating two heterozygous peas would have the probability of producing 25 percent of its progeny with homozygous yellow genes, 50 percent with heterozygous color genes, and 25 percent with homozygous green genes; figure 10 illustrates this. Both the homozygous yellow and heterozygous progeny would be colored yellow, so on average three-quarters of all of this second generation would be yellow. The 25 percent of the progeny that are homozygous green would be green in color.

FIGURE 10: PEA COLOR

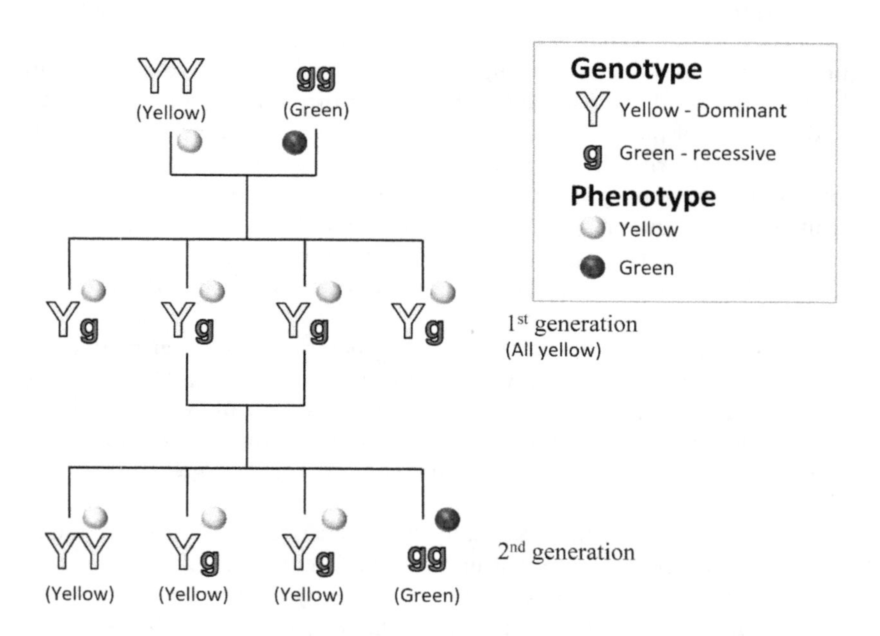

The color of the pea is called its phenotype. The phenotype is what we can observe or measure (such as a biochemical test) about an organism without directly looking at the genes. The genetic content of the pea is called its genotype. That is, a pea that is heterozygous with regard to the color gene (YG in figure 10) would have a phenotype of yellow but a mixed genotype (YG). A homozygous yellow pea would have a genotype of YY in figure 10 but would have the same yellow phenotype as a heterozygous yellow pea. The genotype of a phenotypically green pea would always be GG. Nowhere in Mendel's garden was a progeny that was greenish yellow in color, thus debunking the blending theory of inheritance. Also, nothing that happened to the peas during their lifetime affected the mathematics regarding the probabilities of the future generations. Thus Lamarckian evolution was also debunked.

We have all heard this same lesson with regard to people with blue eyes and brown eyes. We learned that "big B" was the gene for brown eyes and "little b" was the gene for blue eyes. We also learned that "big B" was dominant. Thus if you had the genotype BB or Bb, you would have brown eyes—that is, your phenotype would be brown eyes. If you had the genotype bb, then you would have blue eyes, and therefore your phenotype would be blue eyes. If both of your parents had the genotype Bb, then you would have one chance in four of having blue eyes even though both of your parents had brown eyes.

What Mendel also showed was that different traits are independently inherited. We now call the two variants of the gene that codes for pea color *alleles* for that gene. Thus, there is a yellow allele for the pea color gene and a green allele, and one allele is inherited from each parent. Mendel did the same experiment with other pea traits. For example, some peas have smooth skin texture, and some have wrinkled skin texture, with smooth being dominant over wrinkled. If one were to diagram the first two generations regarding skin texture, it would look identical to figure 10 if you substitute an S (smooth) for the Y and a W (wrinkled) for the G. The skin texture allele is different from and independent of the skin color allele. Because peas contain both genes for color and skin texture, when Mendel plotted the mathematics of the various combinations for the progeny, he observed that each of the traits independently maintains its inheritance pattern. Thus you could have any combination of genotypes among the two genes, as well as any combination of resulting phenotypes, and they would be in the same mathematical proportion as predicted by their independent inheritance.

Figure 11 demonstrates this. It shows two generations after mating of a homozygous yellow smooth pea with a homozygous green wrinkled pea. Note that in the first generation, all the progeny are phenotypically yellow and smooth. However, in the second generation, there are nine yellow and smooth, three yellow and wrinkled, three green and smooth, and one green and wrinkled.

I realize that figure 11 will make many eyes glaze over so I just want you to take away this one point. Note that in the second generation, 25 percent of the progeny are green, and 25 percent of the progeny are wrinkled—the same proportions as would be predicted from the single gene matings. These genes are acting independently!

Figure 11: Combination Pea Color and Texture

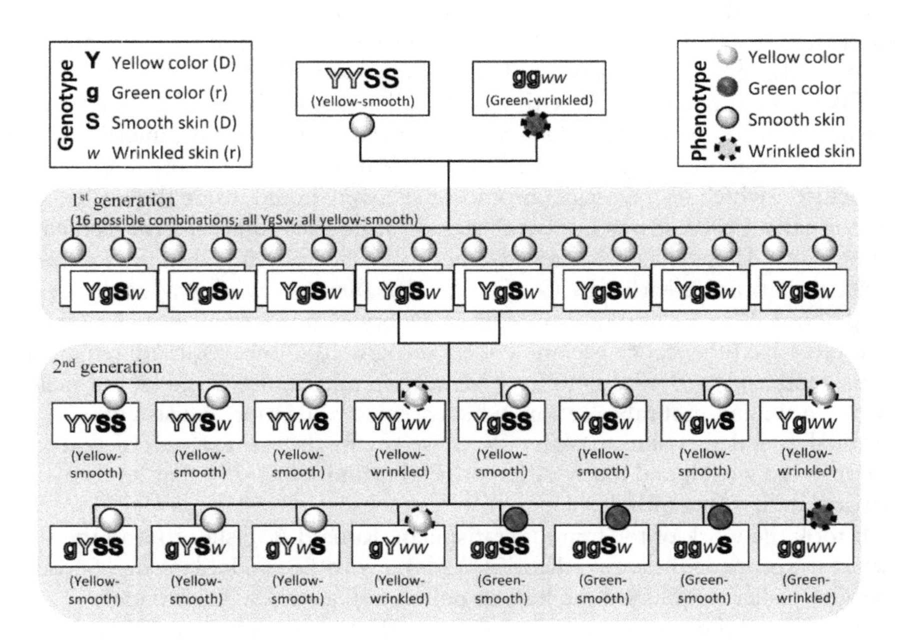

In retrospect, twentieth-century statisticians reviewed Mendel's publications and concluded that his results might have been too good to be true. That is, they seemed fudged to prove Mendel's theories rather than reflect the normal range of error and variation usually found in natural experiments. Statistically speaking, mating two hybrids would not always give the exact ratio of three to one in offspring phenotypes as Mendel's data showed. Rather, random chance would create various results around those values.[62]

It isn't as easy as Mendel's peas.

Nonetheless, Mendel's ratios were correct, and we still use the term Mendelian inheritance. Further, the concept of alleles consisting of one gene variant inherited from each parent is correct. However, the way genes really work is far more complicated than Mendel's peas. For example, two blue-eyed parents legitimately can have a brown-eyed child, which seems to contradict Mendel's theory (see appendix 4-B). Sometimes different genes are statistically linked to each other in inheritance, which isn't explained by Mendel's theory. A gene variant can be both dominant and recessive, which doesn't seem compatible with Mendel's theory. The thyroid gland has all the genes necessary to produce insulin, but it doesn't (only the pancreas makes insulin), and Mendel's theory doesn't explain that. The butterfly and caterpillar shown in figure 7 have identical genes. Identical twins do not always have identical genomes. Some traits sometimes appear to be blends of the parents like skin color and hair texture. What exactly is a gene? And how does all of this impact our discussion of *Homo nouveau*? There are answers to these questions and many more when you read on! Although we will dig deeper into genetics, I will keep this discussion relevant to our search for a path to *Homo nouveau*. I have moved some of the less relevant aspects of genetics to appendix 4.

Taking a closer look at these things we call genes, in chapter 2 I stated that we are in the domain called Eukarya. That means our cells have nuclei that are separated from the body of the cell by a nuclear membrane. In the late 1800s, structures (which we now call chromosomes) were noted inside the nuclei, and by the early 1900s, it was known that these structures contained the elements of heredity. These elements, named genes in 1909, were linked to the inheritance traits studied earlier by Gregor Mendel. Mendel's work was re-discovered about this time and the concepts of genotype, phenotype, and dominant and recessive inheritance became widely recognized. However, there remained considerable controversy as to how Mendelian inheritance could be reconciled with Darwinian natural selection, which seemed to be more consistent with gradual, blended evolution. After all, we see many different shades of skin color in the world, and many different heights and many different levels of artistic ability for example.

It took the work of the famous statistician, Ronald A. Fisher to lead the effort to reconcile these two theories.[63] Along with J.B.S. Haldane and Sewall Wright, Fisher established the field of population genetics. Their papers demonstrated the mathematical support for the mechanism of natural selection based on Mendelian inheritance. They showed how Mendelian genetics was efficient in preserving the variation within species that allowed natural selection to create the gradual evolutionary transitions found in paleontology. With the later help of other prominent geneticists including Theodosius Dobzhansky and Ernst Mayr, this reconciliation became known as the modern evolutionary synthesis. This grand synthesis brought together key concepts of Darwinian evolution, Mendelian genetics, genetic drift and paleontology.

How does it all work?

It was later determined that the chromosomes in the nuclei were made up of long strings of a molecule called DNA (deoxyribonucleic acid). By the late 1940s, we knew that humans had twenty-three pairs of chromosomes, or forty-six total. One of these twenty-three pairs determines sex (labeled X for female and Y for male). The other twenty-two pairs are called autosomal chromosomes. *Autosomal* simply refers to the chromosomes that are not sex chromosomes.

In 1953, Watson and Crick published their Nobel Prize–winning study on the double helix configuration of DNA.[64] A helix is a spiral, and a double helix is two spirals intertwined. In their paper, Watson and Crick stated, "It has not escaped our notice that the specific pairing we have postulated immediately suggests a possible copying mechanism for the genetic material." What an understatement! That "possible copying mechanism" revolutionized our understanding of genes.

From the Human Genome Project, we learned that *Homo sapiens* has between twenty and twenty-five thousand different genes. Genes are segments of DNA contained in our twenty-three pairs of chromosomes. Each chromosome contains hundreds or thousands of genes depending on the chromosome. Each chromosome consists of two long strands of DNA that are bound together in the famous double helix configuration. A gene is a piece of that DNA sequence that determines a specific feature, function, or trait of the organism. There are two copies of each gene on the chromosome pair consistent with Mendel's observations, one on each of the corresponding locations on each pair of chromosomes. These are the alleles, discussed earlier. In sexually reproducing animals, one chromosome of each pair, and therefore one variant of each allele, comes from the maternal side (or egg), and one comes from the paternal side (or sperm).

DNA itself is a sequence of four different types of compounds called *nucleotides*. Human DNA contains over three billion pairs of nucleotides in its chromosomes, and the entire sequence of these nucleotides is called the *genome*. The purpose of a gene is to produce a protein. (See box for how a gene works.)

Thus the more than twenty thousand genes in the human genome really make proteins. These are referred to as the protein-coding genes, or simply coding genes. Proteins do most of the things in our body that make us function. That is true of all plants and animals. There is no life without proteins. Even the simplest living organisms like bacteria require a minimum of several hundred protein-coding genes.[65] They are what gave Mendel's peas their color and texture. Some proteins are structural to form various types of cells, tissues, and organs; others are functional, like enzymes that cause various chemical reactions to occur in our body to create different types of compounds other than proteins.

Knowing that the function of genes is to create proteins helps one understand the meaning of dominant and recessive genes. It is not as though the dominant gene does something to control the recessive gene (i.e., in some way to dominate it). In most cases, it simply means that the dominant variant of the gene produces a protein, and either the recessive variant doesn't produce any protein, or it produces a different protein having no effect or a different

How a gene works

- A gene consists of DNA.
- DNA is used to make messenger RNA (mRNA). RNA consists of four nucleotides, one of which is different from the DNA nucleotides.
- mRNA migrates out of the nucleus into the cytoplasm of the cell and links up with ribosomes.
- The ribosomes contain ribosomal RNA (rRNA)
- mRNA plus rRNA together form a template from which a protein is produced.
- Proteins consist of combinations of 20 amino acids
- Each amino acid is "coded" by one or more specific series of three nucleotides.
- This code is called the codon and is the same in all plants and animals.

effect. See appendix 4-A for a more complete discussion of the meaning of dominant and recessive genes, as well as the special case related to the unique X and Y sex chromosomes.

Before the completion of the Human Genome Project in 2003, it was thought that the human genome contained as many as one hundred thousand genes. It was somewhat of a surprise to learn it was closer to twenty or twenty-five thousand. Because we have as many as one million or more different proteins in our body, if we have at most twenty-five thousand protein-coding genes, where do the other proteins come from? It turns out that they come from these same coding genes through a process called alternative splicing. That is, a single protein-coding gene can produce multiple proteins. Alternative splicing, like everything else in our genome, is complicated, and it is explained in appendix 4-C.

The epigenome

The most remarkable thing we learned from the Human Genome Project was that the coding genes make up only about 1.5 percent of our genome. Further, the twenty thousand or so coding genes were about the same number of genes that exist in other animals—for example, the mouse. What is going on in the other 98.5 percent of our genome? Initially it was thought that the rest of our DNA in the genome did nothing and was labeled junk DNA. However, the more we look at the junk, the more interesting, complicated, and important it turns out to be.

For one, much of the DNA sequences in the junk portion are not just random nucleotides, or some kind of filler material. It turns out that as much as 48 percent of it corresponds to the DNA sequences of ancient viruses that somehow made their way into our DNA over twenty-five million years ago. Although most of this portion of the junk DNA seems to be silent, recently some of it is implicated in triggering autoimmune diseases such as systemic lupus erythematosis.[66 67]

More importantly, most of this junk relates to an entire new world of genetics called epigenetics.[68] The epigenetic portion of our genome is at least as important, or more important than the coding genes. And it is far more complicated. We have learned most of this only recently, since the Human Genome Project was completed. The result is a myriad of publications describing what we now call the epigenome.*

The epigenome is related to the noncoding part of the chromosome. We have learned that the noncoding part of our DNA does a lot of coding—just not for proteins. As described in the box above, in order for the protein-coding portion of DNA to work (what we normally call our genes), it must first produce mRNA to move out of the nucleus and combine with the rRNA to make

*In 2003, the part of the NIH that did the Human Genome Project, the National Human Genome Research Institute, launched a project called ENCODE (Encyclopedia of DNA Elements) to study this other world of DNA. See National Human Genome Research Institute, http://www.genome.gov/10005107.

the proteins. It turns out that the "noncoding" part of our DNA codes for many other kinds of RNA, whose function is not to produce proteins but rather to perform other critical functions. There are many different types of these other RNAs performing these functions (see appendix 4-D). In fact, the DNA that codes for these other RNAs, the so-called noncoding genes, outnumber the coding genes in the human genome.

Nucleotides

- **DNA:** Cytosine (C), Guanine (G), Thymine (T), Adenine (A)
- Double-stranded double helix in nucleus in which C pairs with G, T pairs with A
- **RNA:** Cytosine (C), Guanine (G), Uracil (U), Adenine (A)
- No pairing - single stranded

The functions of these epigenomic RNAs vary widely, are essential for normal growth and metabolism, and play a role as well in disease, including a major role in causing cancer.[69] One of the main functions of these epigenetic RNAs is to regulate the activity and the timing of that activity in the coding genes. By regulate, I mean they either turn on or turn off the production of proteins from a coding gene completely or to moderate the amount of protein produced by a coding gene. The complexity of all of this is mind-boggling, and I'll try to simplify it at the risk of greatly misrepresenting the true processes.

As stated before, DNA is made up of four different building blocks, called nucleotides (see box). The nucleotide sequence constitutes a code that determines the proteins it produces. Proteins are made up of compounds called amino acids, and each code translates into an amino acid. There are twenty different kinds of amino acids used in all plants and animals. The code for each amino acid is a sequence of three specific nucleotides, so every three nucleotides on the coding portion of a gene strand of DNA codes for one amino acid.

Why three nucleotides in the codon?
It is the genius of nature. Because there are twenty different amino acids in proteins, a code of two nucleotides could only account for sixteen of them because there are only sixteen combinations of four nucleotides taken two at a time (4^2). Three is more than enough (4^3, or 64), and four would be overkill.

Using three nucleotides give more combinations than are needed to code for the twenty amino acids used in natural proteins (see box). It turns out that it's important to have more than the minimum twenty combinations that would be needed. Most of the amino acids have more than one code that translates to it. The complete dictionary of these codes is called the codon (see appendix 4–E). Having multiple nucleotide codes for most amino acids makes us less vulnerable to mutations (see below). Also, different codes for the same amino acid affect the timing of the protein construction, which impacts its ultimate three-dimensional shape and function.[70] (See appendix 4-F to learn about the importance of 3-D shapes of DNA and proteins.) Finally, we needed to have some extra codes that don't translate into amino acids in order to signal where the beginning and end of genes are. The coding genes are buried among a lot of noncoding DNA. How things have worked out the way they have in evolution never ceases to amaze me.

Now I will layer on top of this the epigenome. These other RNAs are critical for virtually every function of the genome. They are needed for both mitosis and meiosis to occur normally (see below). They are involved in alternative splicing. They control at least some of the aging processes of cells. They cause the necessary inactivation of one of the X chromosomes in female cells (see appendix 4-G for a more complete explanation of this). Most important, they regulate the activity of the protein-coding genes. One group of them has the effect of preventing or reducing a particular gene or set of genes from producing the coded proteins. They usually do this by binding to the mRNA that that gene produces. They can do this temporarily, or they can permanently alter the DNA chemically in a process called methylation. Methylation doesn't change the DNA nucleotide sequence; rather, it inactivates that coding gene on a permanent basis. This is why a brain cell, for example, doesn't act like a heart cell. Another group of RNAs has the opposite effect; it enhances the ability of a gene to produce protein.

These so-called gene regulators somehow know when to become active and in which cells to become active. This is how the body develops from a single-celled fertilized egg to a complete human. At the beginning, the early cells, called stem cells, can develop into any cell in the body. As the forming fetus begins to differentiate into different types of tissues (e.g., neurological tissue versus intestinal tissue), the cells narrow their potential to develop into a smaller range of related cell types. For example, blood stem cells can develop into either white blood cells or red blood cells, but they cannot normally develop into muscle cells. Ultimately the stem cells further specialize and become the cells of a particular organ or tissue. At each of these stages of development, it is the epigenome that controls this process. It is the conductor of the beautiful and still mysterious symphony of embryology—how the zygote differentiates into a full human. The underlying DNA contained in the nucleus of each cell does not change (except for occasional mutations). A heart muscle cell, for example, still has all the genes in its nucleus to be a neuron or skin cell. However, these genes are all turned off by the epigenome, and only the heart muscle genes are active. The term used to describe this is expression. Only the heart muscle genes are expressed in the heart muscle cells. This variation in gene expression at different times in the life cycle of an organism explains figure 7; a dramatically different subset of the same genome is expressed in the caterpillar phase of a monarch butterfly than is expressed in the butterfly phase. Gene expression controlled by the epigenome is what makes each cell type a specialist.

The epigenome is active from the moment of egg fertilization until death. It does not ever change the DNA nucleotide sequence found in the chromosome. Instead, it directs polypeptides (small proteins) and small pieces of RNA to attach to the DNA or its supporting structures in various ways that alter the expression of the coding genes, including turning them off completely. Some of those changes result in permanent alterations to the shape or chemical composition of certain parts of the DNA (but never the nucleotide sequence), whereas other changes are temporary.

Epigenetic alterations accumulate throughout life. Some of these epigenetic processes are affected by environmental factors such as toxins, diet, smoking, and stress.[71 72] It is only when the germline cells (see below for a description of germline cells) in the testes and ovaries produce a sperm or egg that the slate is wiped clean and the original epigenome is recreated. Although there is uncertainty at this time, there is some evidence that some of the epigenetic alterations that accumulate during life may be passed on in the germline cells.[73] This would resurrect a bit of Lamarckian theory.

Mitochondrial DNA

All of the discussion so far relates to the DNA contained in the nuclear chromosomes, which is by far the major DNA in our cells. There is one other source of DNA in each cell that is contained in small organelles outside the nucleus called mitochondria. Mitochondria are the power plants of our cells that create the energy chemicals needed for our entire metabolism. Mitochondria are thought to have originally been separately functioning bacteria-like organisms that became incorporated into our cells billions of years ago. Compared to nuclear DNA, mitochondrial DNA is miniscule, consisting of only about 16,600 nucleotide pairs (compared to three billion in nuclear DNA) and only thirty-seven genes (compared to twenty thousand–plus nuclear genes). Billions of years ago, the mitochondrial DNA had many more genes, but over the years of evolution, most of them have migrated to the nuclear chromosomes, including some still vital to mitochondrial function, leaving only the small number we have today.[74]

Mitochondrial DNA is double-stranded like nuclear DNA, but rather than being linear, it is a ring. It was the first part of our genome to be sequenced in its entirety and is frequently used to help distinguish eukaryote species. Although mitochondrial DNA is small compared to nuclear DNA, because there are many mitochondria in every cell, which replicate multiple times between each cell division, there are more copies of the mitochondrial DNA to facilitate its analysis. It is inherited only from the mother; the mitochondrial DNA in sperm are either lost or destroyed in the process of egg fertilization. Therefore, mitochondrial DNA is very useful in tracking the maternal lineage. (See appendix 4-H for other information about mitochondria.) The Y chromosome in the nuclear DNA is inherited only from the father and is used to track the paternal lineage. Interestingly, the Y chromosome is our smallest nuclear chromosome.

How does all of this relate to species, speciation and *Homo nouveau*?

I told you this was far more complicated than Mendel's peas. In fact, it is far more complicated than the above description. The very concept or meaning of the word *gene* is becoming fuzzy. I've moved some of this additional complexity into appendix 4 to try to keep this relevant to understanding the answers, when I get that far. I think one take-away from this genetics discussion needs to be simply that it is complicated and much of it is still unknown.

I need to begin relating all of this discussion of genetics back to how species and speciation occur. Every cell in the human body (except mature red blood cells, which don't have nuclei) contains in its nucleus a complete set of the twenty-three chromosome pairs, or forty-six total, one of each pair from the mother and one from the father. As the body grows, new cells are created through a process called mitosis, in which the entire DNA of each chromosome is replicated by copying two additional strands of DNA to form a new copy of the chromosome for the nucleus of a new cell. Even after the body is fully grown, mitosis continues in most tissues to replace worn-out cells or damaged cells. The rate of mitosis and the detailed processes within the cell to initiate and complete a mitotic cell division is controlled by the epigenome and varies from cell type to cell type. We now know that certain mutations in the epigenetic genes can dramatically affect mitotic processes. One of those effects is related to cancer cells becoming out of control.

A couple of billion years ago, a second form of replication evolved called meiosis.[75] In the evolutionary history of life on earth, mitosis came first, and meiosis almost surely evolved from it.[76] In meiosis in humans, specialized cells develop, called germline cells. The male germline cells are contained in the testis, and the female germline cells are contained in the ovary. These cells function differently from all other cells (called somatic cells). They do not replicate by creating a complete new set of chromosome pairs by mitosis; instead, they create sperm in the male and eggs in the female that contain only one of each chromosome pair containing just one copy (allele) of each gene. Each sperm and egg therefore contains only half the amount of chromosomes (twenty-three rather than forty-six) of the parent germline cell. It is only later, when a sperm fertilizes an egg, that the chromosomes get back together and pair up into the full complement of twenty-three pairs in the fertilized egg (called a zygote). Once fertilized, the zygote grows by replication in the mitosis fashion. As the zygote grows into a fetus and finally a complete human being, the cells produced by mitosis are the body cells, or somatic cells. Some of the cells specialize into germline cells in the testis or ovary, and the process starts again.

Crossover recombination

Of course it is more complicated than this. When the germline cell creates a sperm or egg depending on its sex, it does not simply divide the paired chromosomes and put one of them into a sperm or egg. Instead, it goes through a process called crossover recombination, copying genes from one chromosome or the other of each pair in a somewhat random fashion. It usually copies two or more genes in a row from one parental chromosome and then switches over to copy the next genes from the other. It continues this until one copy of every gene allele is made; this includes the coding genes and the noncoding genes or epigenome. Thus the resulting sperm or egg is a true combination of the genes of the two parents but in a very unpredictable way.

Only half of the combined parent genes make it into each sperm or egg, but which half is unpredictable, variable, and always a mixture of gene copies from each parent. In fact, there are so many possible ways any given copy can be made that the resulting sperm and eggs, although consisting of genes from the parents, are quite variable in their genetic makeup from each other. Any given sperm or egg, although containing only half the number of chromosomes of either parent, contains a mixture of genes from both parents. That's why children of the same parents differ, even fraternal twins. After fertilization, the resulting zygote then has a mixture of genes from all four grandparents. Again, the entire meiotic process is under the control of the epigenome.

Because every gene has variations in the population, each of us receives one set of variable genes from the mother and one set from the father. Keep in mind that the regulator DNA, or epigenome, is also undergoing this same random mixing, so the variation is in both the genes and the regulators of those genes. Any given individual will have a different combination of these variations. *Thus, simply by the process of normal mating, genetic variation occurs in a population.* That is, normal sexual reproduction is a constant remixing of the genes and the regulators of those genes. That remixing provides one of the bases for the ultimate development of a new species under the right circumstances. New gene combinations can lead to new phenotypes that can be acted upon by natural selection or create new barriers to interbreeding in the process of speciation.

Mutation

In addition to the normal process of sexual mixing of genes, a second process of genetic change also occurs: mutation. Mutations are random events that alter one or more nucleotides in the DNA and could change a gene. These happen spontaneously, but there are many environmental factors that could increase the rate of such mutations, such as radiation or toxic chemicals. These same environmental factors can also change the epigenome as well. The result of a mutation is unpredictable; it could have no effect on any gene. Some of the DNA in our chromosomes is not involved in genes or their regulation, and so a mutation in these true junk areas has no impact. But even if the mutation occurs in a gene segment, it could have no effect. There are multiple nucleotide codes

Some facts about mutations.

- Approximately 60 germline cell mutations are passed from parents to every child. Most have no effect, but since there are over 200,000 daily births in the world, there is significant genetic change occurring in *Homo sapiens* daily.
- Fathers contribute more than mothers to germline cell mutations with more in older fathers.
- Sometimes entire genes or entire chromosomes are deleted or replicated in excess. These are called errors.
- Errors increase with the age of the mother.
- Mutations in somatic cells occur at a 4-25 times higher rate than in germline cells.

for most amino acids (because we have more combinations than we need for just twenty amino acids), and a single nucleotide change in a gene may not alter the protein it produces at all.* Finally, even if it does alter a single amino acid, that does not necessarily cause any effect on the resulting protein.

The effect of mutation also depends dramatically on whether it occurs in a germline cell or a somatic cell. Germline cell mutations are the more significant because they are passed on to the progeny.[77] But given that each of us has trillions of mitotic cell divisions in somatic cells during our life, the number of mutations that accumulate is staggering.[78] These somatic cell mutations play an important role in cancer and other diseases, as well as aging in the individual, but they are not passed on to progeny.

A mutation could have a negative effect. For example, a single specific change in just one nucleotide changes just one amino acid of one protein and causes sickle cell hemoglobin to be produced. A mutation could cause a beneficial effect. Sometimes it takes many mutations to cause a specific positive or negative effect. For example, some mutations could make a person less susceptible to malaria, or less likely to develop Alzheimer's disease.

The mutation does not have to be at the nucleotide level. One could have a mutation at the entire gene level—for example, the creation of an extra copy of a gene during meiosis or mitosis, or omitting a gene entirely. These are sometimes referred to as errors in gene replication rather than mutations. Errors and mutations explain why identical twins could have nonidentical genomes. Finally, an error can be at the chromosome level. For example, Down's syndrome involves the creation of an extra copy of the chromosome 21, resulting in the individual having three copies rather than two of this chromosome. Again, any one of these errors or mutations does not in itself create new species, but they allow the possibility of new species to evolve. Mutations and meiosis together provide one of the two necessary ingredients for speciation: genetic change. These genetic changes lead to different phenotypes, which are acted upon by natural selection, and they ultimately lead to the development of barriers to interbreeding. That's how new species emerge.

The phenotype is not always strictly determined by the genotype. How the genes interact with the particular environment of the organism also has an impact on the phenotype. Thus, the same genotype may lead to different phenotypes depending on the environment in which the individual lives. For example, two humans could have the same exact genes controlling height and weight. However, if one of the humans is malnourished in childhood, he or she could have stunted growth compared to another well-nourished human. The stunted growth would be the phenotype of that individual.

This adds another variable in the link between genes and speciation. It also complicates the conclusions that can be reached in examining the fossil record,

---

*For example, one DNA code for the amino acid glycine is GGT. A mutation in the third nucleotide (T) to any of the other three nucleotides will still result in one of the codes for glycine (GGG, GGC, GGA). On the other hand, the amino acid phenylalanine has only two DNA codes (TTT, TTC), so a mutation in the third nucleotide has a 66 percent chance in resulting in an amino acid change to Leucine (TTG or TTA). See appendix IV-E for a complete listing of the codon.

which is discussed in the following chapters. Not only must the genotypic variability within a species be considered when looking at the fossil record, but one must also consider the phenotypic variability when examining small numbers of fossils. One cannot generalize accurately about an entire species by looking at either the phenotype or the genotype of a small number of individuals.

I will come back to some of the details of genetics when I consider the various mechanisms that might lead to *Homo nouveau*.

The answers will be related to either genomic or epigenomic changes (or both) from *Homo sapiens.*

The table below outlines the history of the gene.

Year	Event	Who
1866	Inheritance by discreet independent "units"	Gregor Mendel
1870s	Chromatin in nuclei related to inheritance	Walther Flemming
1888	The word *chromosome* is coined	Wilhem von Waldeyer-Hartz
1890–1905	Units of inheritance reside in chromosomes	Theodor Boveri Walter Sutton Nettie Stevens
1900	Mendel's work "rediscovered"	Hugo de Vries Carl Correns Eric von Tschermak-Seysenegg William Bateson
1905	The word *genetics* is coined	William Bateson
1905–1908	Trait unit mapping on chromosome and concept of crossover or recombination	Thomas Morgan Alfred Sturtevant
1909	Units of inheritance called genes. *Phenotype* and *genotype* enter vocabulary	Wilhelm Johannsen
1918	Blended and continuously variable phenotypes explained by multiple discreet genes	Ronald Fisher, T.B.S. Haldane, Sewall Wright
1926	Mutations in fruit flies caused by radiation suggests chemical nature of genes vulnerable to toxins	Hermann Muller
1940s	Environment interacts with genotype to produce phenotype	Theodosius Dobzhansky
1944	Genes consist of DNA	Oswald Avery Colin MacLeod Maclyn McCarty
1945	Genes code for proteins	George Beadle Edward Tatum

Year	Event	Who
1953	DNA is a double helix, suggesting a copying mechanism	Rosalind Franklin James Watson Francis Crick Maurice Wilkins
1959	Genes are regulated by other genes	Arthur Pardee François Jacob Jacques Monad
1960	Messenger RNA is intermediary between genes in the nucleus and protein production in ribosomes in cytoplasm	Sydney Brenner François Jacob
1960s	Coding system from nucleotides to amino acids determined (Codon Table)	George Gamow Francis Crick Sydney Brenner Marshall Nirenberg Heinrich Matthaei
2003	Human Genome Project determines 20,000–25,000 coding genes in humans	Multiple
2000s	Importance of noncoding genes (the epigenome)	Multiple
Future	Genetic engineering	Multiple

Chapter 6

Getting Past the Chimps

"We don't really know for sure how we got here…Somehow, a group of about thirty thousand individuals started to spend more time out of the trees and began an amazing evolutionary journey. They became the first humans."

To prepare for the possibility that *Homo nouveau* will emerge from *Homo sapiens,* it is useful to review how *Homo sapiens* got here. The same or similar mechanisms might occur again, so this history will inform the answers.

We don't really know for sure how we got here. Historically, all of the evidence about species and their ancestral chains came from paleontologists looking at the fossils of extinct animals. Fossils form when a buried organism, under the right conditions, has its fluids and spaces replaced by minerals that take the same form as the body part. This primarily happens to areas of a body that are already mineralized such as bones, teeth, and shells. Softer parts tend to decay away. Sometimes all that is left is an imprint or image of the body part in the surrounding sediment. Finally, sometimes an actual piece of the original organism is preserved (typically a bone or tooth) that hasn't been replaced by mineralization. These are the most valuable fossils because we sometimes can extract their DNA.

It is important to understand that fossil evidence is fragmentary in multiple ways. Most often, it shows only bits and pieces of entire organisms, requiring careful reconstruction to envision the whole. Fossils are also fragmentary in the sense they are from a small number of individuals of any one species. That in itself can be misleading because we know there is great genetic and phenotypic variation within a single species. They are fragmentary in location, with similar fossils found in widely separated regions. Finally, they are fragmentary in time, usually leaving large time gaps between periods in which the discovered fossils were created.

Another factor that distorts and biases fossil evidence is the fact that the environment and geology of a region greatly influences the conditions under which a fossil can even develop and subsequently be discovered. As Robert Foley states in his book *Humans before Humanity*[79] there are areas "that are not necessarily good places to live and to evolve, but good places to die. (Good, that is, if you want to become a fossil.)" These areas have heavy river and lakebed sedimentation, volcanic activity and movement of tectonic plates to cover up and preserve dead animal remains, and tremendous geological upheaval to later uncover the fossils. Therefore, where fossils form and are later discovered may not produce a true picture of evolution but rather reflect the fossils that happen to be in good fossil-producing regions.

In addition to the fossils that are direct physical evidence about a species, other things found next to the fossils also are useful as evidence. Those other things include tools, pottery, evidence of fire use, dwellings, body ornaments, wall art, and even footprints. The fossils of other plants and animals found adjacent to a particular fossil also help date it. The geological stratum (i.e., the rocks and sediment in which the fossil is embedded) is an important time indicator in itself. Finally, the most recent DNA detection tools have been able to extract mitochondrial DNA directly from soil even without any bones or teeth fossils and match that DNA with known extinct species![80]

The impact of rapid DNA sequencing

Notwithstanding the fragmentary nature of the evidence, paleontologists have done an amazing job in reconstructing much of our lineage. In recent years, we have developed the technology to rapidly sequence the entire genomes of any living species, and particularly to be able to do this on large numbers of living organisms around the world. Couple this with the creation of huge genomic data banks and massive increases in computing capability, our knowledge of evolution is accelerating rapidly. (See box for some examples.)

The most amazing DNA development is that in those cases where an actual piece of an extinct organism is sufficiently preserved, we can sometimes perform genomic analyses on extinct species. This is called paleogenomics; more on this later.

Lest we become too smug about our latest tools, there are still many limitations to our current knowledge. The most useful genetic information, that from extinct species, is very limited and very fragmentary. As Fred Smith and James Ahern point out in their book *Origins of Modern Humans*, "it is unlikely that genetics alone will resolve current controversies concerning modern human origins, no matter how much more reliable genetic studies are perceived to be."[81]

Impact of rapid DNA sequencing technology

- Ability to analyze the full genome of any living species and make comparisons
- Ability to sample large numbers of species rather than fragmentary samples
- Sampling of human genomes from all geographic areas to determine timing and location of the spread of Homo sapiens
- Calculation of the characteristics and timing of common ancestors
- Calculation of the timing of the emergence of specific genes and their associated traits
- Confirmation or refutation of fossil evidence
- Paleogenomics: Detection of DNA from ancient fossils or ancient soil

From mammals to humans

To trace our evolution, I'm going to start with mammals. Humans are mammals. That skips about three and a half billion years or so of life on earth prior to the mammals, but I'm only writing one book and I need to get to hominids (the family we are in) in this chapter. Not that those earlier years weren't important, they are just less so with regard to our ability to think about *Homo nouveau.*

Mammals are a class of warm-blooded animals where newborns suckle milk from their mothers' mammary glands and have certain other physical features including a neocortex in the brain. They appeared about 225 million years ago. Most of them have placentas and give birth to live newborns like we do, but some of them, like the platypus, lay eggs, and the newborns hatch outside of the womb and form a subclass of Mammalia.

Figure 12 shows a simplified evolutionary tree from mammals to Hominidae (includes us). That evolution involved all of the principles of genetics and speciation that we've discussed. I want to make it clear in this and subsequent figures that the progressions shown in these figures are meant only to show the lineage of our species. It in no way is meant to show progress implying increasingly successful species on the evolutionary tree. For example, the most successful mammals in terms of numbers, diversity, or ecological niches are rats, bats, and antelopes—certainly not humans.[82] And all mammals pale in comparison to bugs and bacteria!

FIGURE 12

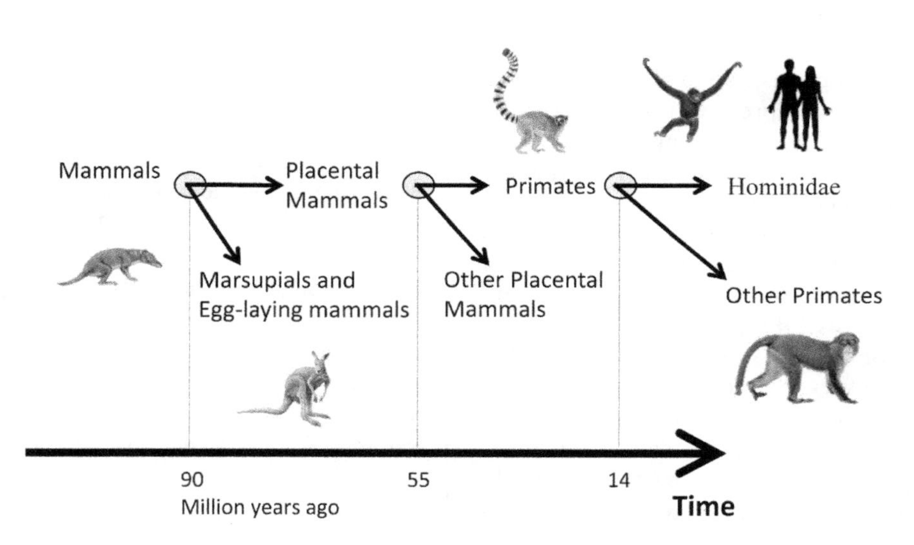

Primates are an order of mammals that include a variety of species well adapted to climbing trees, swinging from branch to branch, but they also walk on two or four legs. They have larger brains in proportion to their size than other mammals. They include lemurs, apes, and humans.

Hominidae, or great apes, consist of the orangutan, gorilla, chimpanzee, and human genuses. Yes, to the dismay of some, we are considered a great ape (although some people use the term to exclude humans). Bonobos are closely related to the chimpanzees and are in the same genus. Orangutans are great apes from Asia. Gorillas, chimpanzees and bonobos live in Africa. Together, we all had a

Timing events: Mammals to Hominidae
(all times are millions of years ago)
225—Mammals first appear
90—Placental mammals split from non-placental
55-85—Primates split off
14—Great apes, including homo genus, emerge
13—Orangutans diverge from other great apes
6.7—Gorillas split off
5-7—Humans diverge from chimps and bonobos
2—Chimps and bonobos split

common primate ancestor until about fourteen million years ago. We don't know exactly what that primate was, but it surely lived somewhere in Africa, probably East Africa.[83]

Figure 13 is a more detailed look at the Hominidae family tree (or hominids). It is pretty well agreed that our closest living relatives are chimpanzees and bonobos. If anything, we are slightly closer to bonobos than chimps. We have many other closer relatives, but they are all extinct.

FIGURE 13

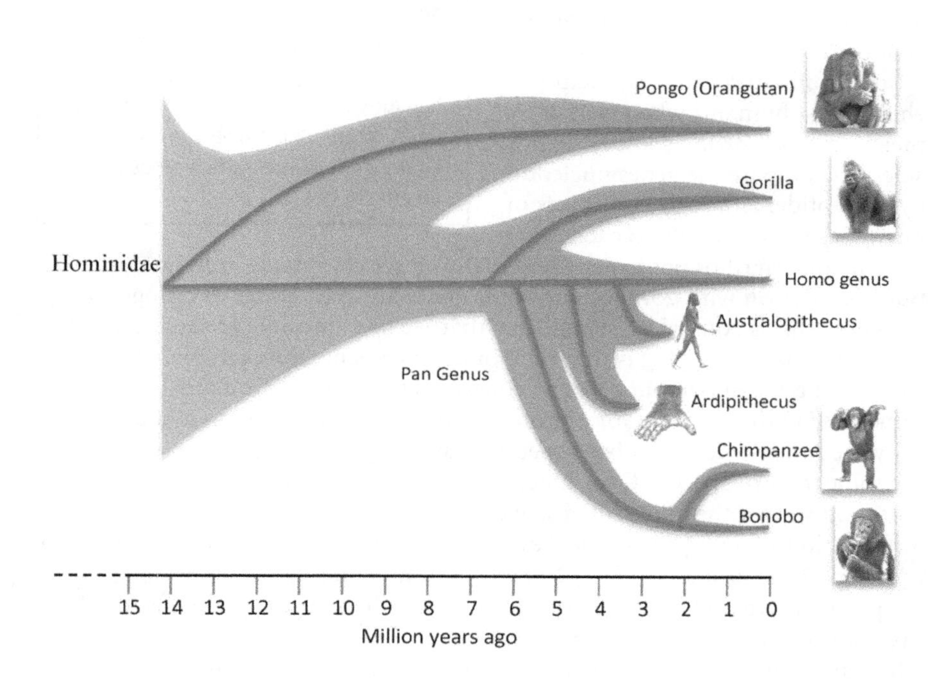

Our closest living relatives

So what exactly does it mean to describe chimps and bonobos as our closest living relatives? Although one can go into great morphologic detail about the shape of the skull, the middle ear bones, the arm bones, and other physical features, the most compelling evidence is in the DNA. Simply put, our genes overlap more with species of chimpanzees and bonobos than any other species. In fact, we overlap in 98.4 percent of our genes. That is, of all our twenty thousand–plus coding genes, chimps and bonobos have the same ones as we do 98.4 percent of the time. Some have even argued that chimps and bonobos are so close to *Homo sapiens* that they should be placed in the *Homo* genus, which would make them humans![84]

As explained by Eugene Harris in his book *Ancestors in Our Genome*, we can even get to a greater level of detail and comparison.[85] We now can look at every gene and every nucleotide within every gene, and can make quantitative statements about the relative closeness of those sequences among different species. When we do this, what is interesting and somewhat surprising is that in spite of our closeness to chimps and bonobos in the big evolutionary picture and at the gene level, they win the closeness contest in only about two-thirds of all of our genes at the micro nucleotide level.

Specifically in comparing today's chimps and humans, when we put their common chromosomes side by side and compare the genes nucleotide by nucleotide, although two-thirds of our genes match closely to each other, roughly one-third of our genes are actually closer on a nucleotide-by-nucleotide comparison with gorillas, and some genes are even closer to orangutans.[86]

Number of chromosomes
Chimps, bonobos, gorillas, and orangutans have twenty-four pairs of chromosomes compared to our twenty-three pairs. It appears that in the course of evolution, two of their shorter pairs of chromosomes merged into one when humans emerged. Nonetheless, their chromosomes are very similar to ours. For comparison, the red-fronted lemur has thirty pairs of chromosomes, a dog has thirty-nine pairs, a pigeon has forty pairs, and a snail has twelve pairs. You don't learn very much by knowing the number of chromosomes an animal has.

I know this is confusing. When I say that we have mostly the same genes as a chimp, what I'm saying is that each of us has a gene, for example, that codes for one of the hemoglobin proteins contained in our red blood cells. It is the same gene in that it codes for a protein that has the same function and basically the same structure in both species. But that protein is not always *exactly* the same in both species. Over time, some of the nucleotides have changed by mutation so that the amino acid sequence of some proteins may be slightly different in the two species, but they're not different enough that it changes the function of the protein. In fact, 20 percent of human and chimp proteins are identical, and the other 80 percent differ by only one or two amino acids. This raises the question as to why there is so much difference between a chimp and a human if our genes and proteins are near identical. The differences are largely in the epigenome.[87] That's the same reason that the same genome produces a butterfly and a caterpillar, or a brain cell and a liver cell.

In summary, in comparing a chimp to a human, 98.4 percent of all of our protein-coding genes do the same thing.* Two-thirds of these genes are more alike at the nucleotide level than between humans and any other living species, which makes chimps our closest relative. Gorillas would be a close second.[88] We do not yet have the ability to compare the epigenome in the same quantitative fashion.

What if we did the same comparison of gene nucleotides of two different individual humans today? First of all, 100 percent of the protein-coding genes would be the same rather than just 98.4 percent. At the nucleotide level,

*There are other methods than this nucleotide comparison of coding genes to compare the genomes of two species that show as much as 5 percent difference between chimps and humans (see R. Britten, PNAS 99:13633, 2002). Nonetheless, chimps and bonobos remain the closest relatives.

there would be differences between any two humans today, but the differences would be about one-tenth as frequent as we see between humans and chimps.[89] That's comforting to know.

What does the nucleotide analysis tell us? As discussed in chapter 5, our genes undergo constant change through meiosis and mutation. But not all genes change at the same time or at the same rate. These changes accumulate in any given species over time. Thus at any given time, we may have genes that are identical or similar to genes in any one of our ancestral lineage species, going all the way back to single-celled animals. In fact, humans have many genes that are identical or very similar to genes in bacteria! One reason for this is that our mitochondria, which have their own set of genes inherited from the female line, originally came from bacteria being incorporated into eukaryote cells. But we also have similar genes to bacteria in our nuclear DNA as well. The main point is we have many more genes identical or similar to chimps than any other species. See appendix 8 for a more detailed explanation of gene comparisons.

What's the point?

What is the takeaway from all of this discussion regarding our closest ancestor as it relates to the answers? First, it is remarkable to me how little change there is between the genomes of dramatically different species belonging to not only different genera but also different families of organisms.

It tells me that we don't need much change to get to *Homo nouveau,* which will be a species within our same genus. It also suggests to me that we need to look closely at the epigenome.

The discussion needs to move on beyond the chimps and narrow this down even further. Now, thanks to both the paleontological and DNA evidence, we know that hominids emerged about fourteen million years ago, and the immediate precursors to our genus split from the chimps much more recently. Specifically, there was a time about 5.4 million years ago when an ancient precursor to *Homo sapiens* looked somewhat like today's chimps. Like humans, the chimps have also evolved from this common ancestor, so it certainly was also different from today's chimps. This common ancestor was a hairy creature that lived somewhere in East Africa, and it spent most of the time hanging out in trees but was able to walk on its two hind legs when on the ground. The weather was hot. Its brain was larger than other primates but still much smaller than ours. It could not speak, use tools, or make and control fire. The males had large testes in order to produce lots of very competitive sperm to compete with other males vying to pass their genes on to the next generation through the very promiscuous females.

Somehow in this milieu, a group of about thirty thousand individuals[90] started to spend more time out of the trees and began an amazing evolutionary journey. They became the first humans.

Chapter 7
Getting to Humans

"We have the only brain that can even ask the question about brains."

I said in the last chapter that we really don't know for sure how we got here. The story gets confusing when we try to piece together the direct lineage of the *Homo* (human) genus. Ian Tattersall is a paleoanthropologist and curator emeritus of the American Museum of Natural History in New York City. He has studied and written extensively on the origins of *Homo sapiens*. His book *The Strange Case of the Rickety Cossack* is both educational and entertaining.[91] He reviews in great detail the numerous twists and turns in trying to piece together the fossil evidence related to our origins. The story is neither pretty nor clear. It is a tale that involves almost as much variation in the backgrounds of the scientists as it does in the science being studied. We have paleoanthropologists, paleontologists, systematists, anatomists, evolutionary psychologists, cultural anthropologists, neuroanatomists, taxonomists, and paleogenomicists vying for attention. That's enough "ists", "cists" and "ologists" to confuse anyone.

There were many conflicting reports based on fragments of bones found in caves, sediments, and ancient seabeds all over the world. These differences sometimes led to petty personal conflicts, shouting matches at professional meetings, and even lifelong feuds. There was claim jumping at archeological sites, political intrigue at the highest levels of governments, the Piltdown man fraud, and lawsuits. There were lots of Leakeys. Theories of our origin ranged from narrow, straight-line evolutionary paths to complex matrices leading to multiple independent lineages reflected in our current cultural and geographic diversity. Finally, the list of purported genuses and species in the *Homo* lineage is as long and sometimes as ephemeral in the fossil record as they are in the *Homo sapiens'* literature (see appendix 6). The following synthesis in this and the next chapter is my interpretation of all of this. It is surely only one of many possible reconstructions of our history, but it's representative enough to inform our search for the answers in chapter 14.

The first inklings of humans

After the chimps split off from the other hominids 5.4 million years ago, our ancient ancestors moved out of the trees and eventually away from the forests of Africa, onto its more open plains. We developed a more upright posture and a bipedal gait that accommodated long walks and running for hunting and gathering. We lost most of our body hair. Somewhere along the way, our brains got bigger and developed the ability to make the first tools, used primarily to carve up scavenged game.

All of these changes have raised many questions and much debate. Why did we lose our protective hair if we were going to be more exposed to sunlight? Why didn't the chimps also develop big brains? Why a bipedal gait? Which came first: big brains or bipedal gait?

Let's start with the first question. One of the things that we know happened between our chimp-like* precursor and humans is that we lost most of our body hair and developed the ability to sweat. Chimps and other primates have full body hair and don't sweat. They also have light skin. But if the premise is true that this happened in the African savannas two to five million years ago, a hairless human would get pretty sunburned without the protection of a dark skin and would get pretty overheated without the ability to sweat. The earliest humans didn't know how to make clothes or sunblock.

This potential problem gave our clever *Homo sapiens* of today a wonderful clue. We must have developed some genes early on that led to dark skin and sweat glands. Indeed, there is at least one such gene that produces a substance that turns on the production of melanin in skin (melanin is the pigment in skin that makes us dark and protects us from ultraviolet radiation). The substance is called melanocortin 1 receptor protein (MC1R). The presence or absence of MC1R is largely the difference between dark skin and light skin. By studying the gene for MC1R, we have learned a great deal about our early ancestors.

It turns out that all current day dark-skinned Africans have this gene, and in fact there is no genetic variation in it. That is pretty amazing given that there has been plenty of time for mutations to occur. The likely answer is that mutations did occur, but because of the high survival importance of having dark skin in the hot African environment, any mutations that reduced MC1R's ability to function would have been quickly wiped out by negative natural selection. Looking at Asians and Europeans, the MC1R gene has lots of variations, many of which make it ineffective. That's okay because Asians and Europeans can live just fine with lighter skin. In fact, reduced MC1R was probably selected as humans moved north to compensate for the lower amount of sunlight. After all, we need sunlight to make vitamin D to keep our bones strong. Likewise, chimpanzees, which live in Africa, also have variability and reduced effectiveness in the MC1R gene, which is also okay because they have all that protective hair to cover their light skin.

By using genetic dating techniques, the gene for the MC1R gene was clearly operating by 1.2 million years ago.[92] This means that sometime between 5.4 million and 1.2 million years ago, the early human ancestors lost most of their hair but were protected by the MC1R gene. There is evidence that sweat glands probably increased with the loss of hair follicles during this same period.[93] This hairless, sweaty human was probably *Homo erectus*, but we'll get to that later.

One question is why natural selection caused us to lose most of our body hair in the first place, if we had to evolve dark skin to compensate for it. The speculated answer is twofold. First, in moving from the shady environment of trees to the open, sunny savannas, full body hair is actually a liability because the body would overheat; it would be like wearing an overcoat in the summer. Losing body hair allowed air to flow freely over the skin and, with the combination of sweat glands, allow heat dispersal. Upright posture further added to the ability to cool off. The speculation is that with our new lifestyle of foraging and running around the savanna, we needed to develop stronger joints and

---

* "Chimp-like" is really a misnomer because today's chimps have evolved away from this common ancestor as well. The common ancestor is probably closer to today's chimps than humans because the chimp's habitats have not changed as much. The phrase should really be "common ancestor with the chimp" but that's too cumbersome, so I'll use "chimp-like" for short.

bones. That requires more vitamin D to be produced with the help of sunlight. Our evolutionary compromise allowed enough of the sunlight in to make vitamin D (by losing hair) but kept out enough of the damaging UV wavelengths (by adding melanin). Pretty clever.

Our brains are the big deal.

What we really need to know is how and why the human brain developed from 5.4 million years ago, when we were chimp-like and had a brain one-third our size. Sorting that out has been a very complicated task for us. Not only is the structure and function of the modern brain complex, but it is largely not understood today. Our brain is not the largest brain of any animal; whale brains and elephant brains are bigger. It is not even the largest brain relative to body mass—a marmoset's brain is bigger in that regard.[94] What differentiates us is that we have more neurons in our brain than other animals.[95] It turns out that brain size alone does not determine how many neurons a species has; one must also consider the density of neurons. For example, it has been observed that some birds, particularly crows, can create and use tools, plan ahead, and perform other cognitive feats that seem far superior to that expected from the size of their brains. Studies of these birds have shown that they pack twice as many neurons per cubic inch into their brains than mammals.[96] *Homo sapiens* not only have large brains relative to our body size, but we also have neuron-dense brains. Finally, the organization of these neurons is probably the key to our intelligence.

We have the only brain that can even ask the question about brains. Our cognitive ability is unequaled and has allowed us to dominate in all of our ecological niches. That most definitely has as much to do with the way the brain is wired as it has to do with size and neuron count.[97] The rapid explosion in population of *Homo sapiens* in the past fifty thousand years attests to our reproductive success. Our brain has been selected by nature to allow us to accomplish this.

Why didn't the chimps also develop a large brain?

Having a large brain comes with a cost; otherwise, lots of other species would likely have evolved bigger brains too. It is the largest energy-consuming part of the body, consuming 25 percent of our calories when we're at rest compared to 8 percent for an ape.[98] Thus our total energy consumption per day is likewise greater than an ape.[99] That means there had to have been genetic mutations simultaneously causing other changes to occur in our chemistry and metabolism to be able to support this monster energy-consuming organ. Evolution was at a branch point: either go with larger brains but also compensate for all the burdens that it brings, or take an easier, smaller-brain path. Evolution doesn't make conscious choices; the genetic changes are random, and natural selection and other environmental forces then take over. Somehow, we lucked out. The big brain experiment happened to occur in our lineage, and fortunately for us, it worked out. With our more intelligent brain, we learned to cook food by controlling fire, which allowed us to ingest far more calories per meal than the old diet of bone marrow, raw meat, and raw plants. Cooked food is easier to chew and digest; that's why we have smaller teeth, jaws, and masseter muscles than the apes. It also allowed us to evolve smaller digestive organs relative to the apes, thus reducing the energy consumption required to digest food.[100] We are fatter than apes, giving us some backup in case food is scarce. We also got smarter at finding food, developed better tools to kill game, and finally domesticated plants and animals—all to better feed this big, intelligent brain with our smaller, weaker muscles. It fit together for us—luckily.

We also know that in infancy and childhood, the current human brain takes a lot longer to mature than any other species. The limitations in the size of the female birth canal would require that if a big brain is a key to survival and superiority, of necessity it would either have to get big after birth, or females would need to evolve bigger birth canals. Also, large brains during gestation would sap too much energy from the pregnant mother. For whatever reason (again, probably random), evolution took the growth after birth path. This has a number of consequences. It allows for a much longer learning phase for language, skills of all types, and cultural development. It also makes offspring much more dependent on parents for a longer period of time. Although that dependency has some negative consequences in terms of the human child's ability to fend for itself, it has the positive effect of spreading out the time to raise children. Humans raise more children simultaneously than apes. Apes generally have only one offspring every four to five years. Is that because prolonging the time to raise human children makes it easier to have them more frequently? Or did we start have children more frequently before our brains grew? That isn't clear, and we still have much to learn about our evolution. So where are we in our understanding?

Ardi

An early step on the path from chimp-like to human was revealed in the discovery of parts of a skeleton in Ethiopia, which has been named *Ardipithecus ramidus* (see figure 13 in the previous chapter). Ardi, as this female is called, is 4.4 million years old. Her foot is chimp-like in that it has an opposable great toe for use in the trees, but she could plant her feet flatly on the ground, did not "knuckle walk," and had other skeletal features indicating an upright posture. Therefore, she could climb trees (more slowly than chimps) but also walked on two legs (but not as far as Lucy, discussed below). She probably spent most of her time on the ground but still climbed trees for food and safety. Her brain was chimp sized. This indicates that upright posture and the ability to navigate bipedally on the ground preceded the improvements in our brains[101] and also occurred while we were still in the woods, not on the plains.[102]

Lucy

After Ardi came Lucy, evolutionarily speaking. This discovery of parts of a female skeleton, also in Ethiopia, was one of the most heralded fossil discoveries of all time. Named *Australopithecus* (Southern Man) *afarensis*, its discovery preceded the discovery of Ardi. At the time, it was often referred to as the missing link between apes and humans.[103] We now know that there are many other links and possible links, including *Australopithecus anamensis* (an older fossil than Lucy) and Ardi, but Lucy characterizes another major shift in our history. She lived 3.2 million years ago (figure 13). She was small, between three and four feet in height fully grown. By then, she had lost her opposing big toe and clearly spent some time in trees and but mostly walking upright beneath the trees. Her brain was still relatively small, and her abilities were more like an ape than a human. She is probably the common ancestor to a number of later australopithecine species and, most importantly, the *Homo* genus.[104]

When I say she was probably the common ancestor to the *Homo* genus, there is really too fragmentary fossil evidence to know for sure the exact lineage in that period between 5.4 million and 2 million years ago. Figure 13 is drawn as though both Ardi and Lucy represent side evolutionary species from the primary *Homo* lineage. Figure 14 shows a blown-up view of this portion of figure 13. Version 1 is the same as figure 13. Version 2 would assume that Lucy was in the direct line of evolution of the *Homo* genus but Ardi was not; in this case, *Australopithecus anamensis* would be the predecessor to Lucy and the *Homo* genus, but not Ardi.[105] Finally, version 3 shows Ardi in the direct line to Lucy, who in turn was in the direct line to *Homo*. We don't have any direct genomic information from this period, so resolution of this lineage question (if it ever is going to happen) must await more complete fossil and genomic evidence.

FIGURE 14

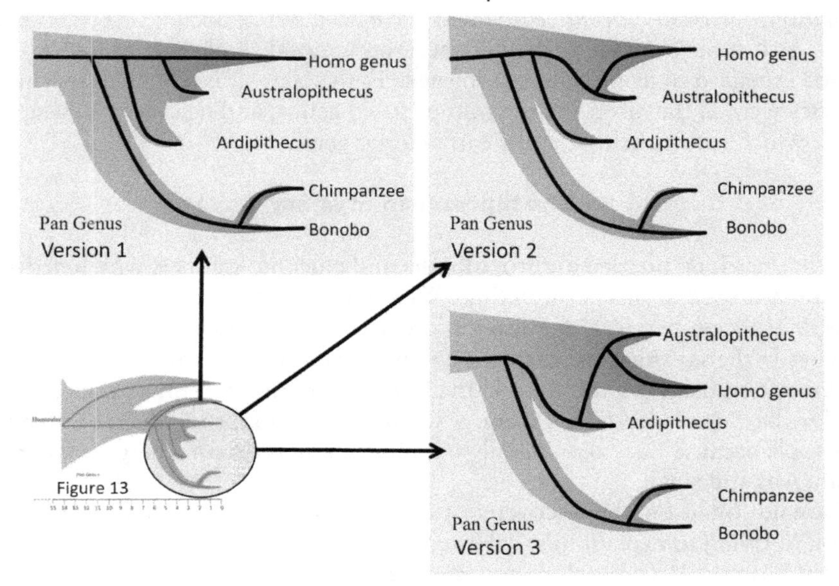

There are those who would argue none of the alternatives depicted in figure 14 is correct. There are fossils that are arguably in the direct *Homo* ancestry that are older than Ardi and have been discovered since she was discovered. These go by such names as *Ardipithecus kadabba, Orrorin tugensis,* and *Sahelanthropis tchadensis.* They date to as long ago as seven million years, and if any of them emerges as the winner as the oldest pre-*Homo* hominid, lots of figures will need to be redrawn. The raging debate centers around which one walked more like a human.[106]

The difficulty in determining our lineage

The problem with determining which fossils belong to species in our direct lineage versus those that are dead-end offshoots goes back to the fragmentation issue discussed earlier. Rarely does one find more than bits and pieces of a single individual in the fossil record. When paleontologists find an apelike jaw, the natural assumption is that the rest of the body would also be apelike, or if a *Homo*-like ankle is found, that it is part of a *Homo*-like creature. But evolution doesn't always work in a coordinated fashion where all parts of the body change at the same rate or even in the same direction.

This was illustrated beautifully when a rare fossil find of virtually entire skeletons of multiple individuals of species *Australopithecus sediba* was discovered in South Africa.[107] This species has a wide pelvis suggestive of the *Homo* lineage, which was thought to have evolved to accommodate larger brains, yet it has a small skull indicating a brain the size of chimps. It has an ankle that appears *Homo*-like but a heel bone that is apelike. It has a long arm like an ape but short fingers thought to be useful for using tools—except it didn't have tools. Depending on which bit of the skeleton might have been discovered in isolation, *Australopithecus sediba* may have been put in our lineage or outside of it. As it turns out, this particular species is a mosaic and remains highly debated in regard to its place in our lineage. To this date, it is still unclear which species were in the direct lineage to the *Homo* genus, and neither is it even clear which species belong in the early *Homo* genus.[108]

The bipedalism mystery

What has long puzzled anthropologists and paleontologists is why bipedalism developed at this point in time. Why did Lucy and the older pre-*Homo* species walk on two legs? It wasn't in order to be able to see over the tall grasses in the savannas (as has been commonly thought) because even as late as Lucy, our ancestors still lived in the forests, spent a lot of time in the trees and walked around between them. It wasn't to free up the hands to be able to use tools because Lucy didn't really use tools any more than chimps do; her brain was still small.

Donald Johanson, the discoverer of Lucy, devotes an entire chapter in his book to trying to explain this.[109] His conclusion is based on the work of C. Owen Lovejoy.[110] The explanation is somewhat complex and difficult to understand, but it goes something like this: bipedalism freed the Australopithecus mother to be able to hold onto one or two offspring while still becoming pregnant with another, thus helping overcome the limitation in numbers of children she could have. Great apes have a child only once every four or five years. Freeing of the mother's hands was needed because the infant hand and foot couldn't cling by itself to the mother like a chimp baby does. *Homo sapiens* is the only species today in which parents hold hands with their offspring while walking to protect them from all sorts of dangers—and they continue to do this long after the offspring are able to walk. Our ancestors probably did the same.

Having free hands also enabled the Australopithecus father to carry back more high-energy food from his foraging outings. All of this led to longer term bonding between a male and female toward the evolution of the human nuclear family. After all, a male wouldn't risk going away for long periods of time to find food if his female didn't remain committed. I'm not sure I'm buying all of this, but it's one theory.

An interesting side note in the Lovejoy article cited above is the meticulous study of the fossil anatomy demonstrating the adaptations of the Australopithecus pelvis and other parts that enabled upright posture. Lovejoy makes the case that the later evolution of the large brain and related pelvic changes to accommodate it actually made the human pelvis and back <u>less</u> well adapted to upright posture. That is, Lucy was better adapted than we are to bipedality.

Another possible explanation for bipedality is that big cats are less likely to prey on upright humans than bent-over humans. This one comes courtesy of circus lore: if a lion or tiger trainer drops his whip, he never bends down to pick it up.[111] Jablonski and Chaplin proposed another explanation. Our cousin apes rise up on their hind legs as signs of aggression, dominance, and threat. Charging with arms held high, arm waving and holding potential weapons like branches, often accompany this. This behavior usually results in one of the competitors backing off or showing some sign of appeasement rather than physical violence. This bipedal mechanism of resolving territorial or mating conflicts saves considerable morbidity and mortality from actual violent encounters. This would therefore have a selective survival advantage.[112]

McHenry attempts to explain it in yet another way. The natural habitat for the apes is trees. In sub-Saharan Africa during the period when bipedalism developed, climate change was thinning out the forests, requiring the tree-dwellers to move increasingly longer distances on the ground between trees to find food.[113] But why bipedalism to move long distances rather than quadrupedalism? Certainly many mammals are quite capable of moving long distances on the ground using four legs. Peter Wheeler has a theory for this. With less shade and more walking between trees, early hominids became more exposed to sunlight and overheating. Upright posture decreases overall sun exposure and allows better air circulation and cooling.[114]

Robert Foley makes an interesting observation to explain bipedalism. Another primate genus, the non-hominid baboons, were also forced from the trees to become ground dwellers about this time and, like hominids, were quite successful in making the transition. However they are fully quadrupedal. Why the difference? Baboons are descendants of monkeys. Monkeys navigate in trees by walking along on top of branches on all four limbs. Hominids are descendants of apes. Apes navigate in trees vertically by hanging from branches. Thus it was natural for hominids to retain this upright posture when transitioning to the ground whereas it was natural for baboons to remain quadrupeds.[115]

Finally, and to me one of the more convincing arguments, if true, is that bipedal locomotion is significantly more energy efficient for hominid mammals than quadrupedal locomotion.[116] With the need to travel longer distances for food as the forests thinned, the energy efficiency of bipedal gait would have been an important survival mechanism. Although non-hominid mammals also move efficiently, it was important for the hominids to retain their upper limb capability to climb trees, carry food and other functions. C. Owen Lovejoy has challenged this theory claiming that the energy efficiency studies were flawed and that bipedal gait is not more energy efficient[117] nor was energy efficiency a factor leading to bipedality.[118]

Obviously we don't really know the answer to the *why* of the bipedalism-timing question. We do know the *when*. It was while we still lived in the forests and climbed trees a lot, and before we got big brains.

I like the answer supposedly given by a Tibetan scholar when asked why we are bipedal: "A sense of humor."[119]

Brain growth kicks in and we became humans.

It wasn't until about two million years ago that our brains finally started to take over. *Homo habilis* (handy man), arguably the first human species, was so named because it clearly used some of the first stone tools. These flakes of rocks were thought to be used for scavenging bones rather than hunting. I say arguably the first human species because some paleontologists think that *Homo ergaster* should get the privilege of being the first humans, and *Homo habilis* should more properly be called *Australopithecus habilis*. Other than the primitive tool use, there is not much to distinguish *Homo habilis* from *Australopithecus* in terms of morphology, habits, or other characteristics.[120] However this uncertainly gets resolved over time, clearly our brains began to grow in size and capability between 2.0 and 1.5 million years ago, with new species branching from our main evolutionary line and culminating in *Homo erectus*. All of this happened in Eastern and sub-Saharan Africa.

The control of fire is a major factor in the evolution of the *Homo* genus. In the transition from Ardi to Lucy to the earliest humans, less time was spent in trees and more time on the ground, culminating in complete ground dwelling. Sleeping on the ground was dangerous, exposing these early humans to carnivorous predators. Sleeping near a fire would have provided some protection. Probably this began with the Habilines (*Homo habilis)* or soon thereafter. The more significant use of fire came later, when it was used for cooking meat and other raw foodstuffs. This allowed greater caloric intake with less caloric cost of digestion—a requirement to support the ever-enlarging brain. Exactly when cooking became a part of the *Homo* culture is not known, but by eight hundred thousand years ago (before *Homo sapiens*) it was well established.[121]

Homo ergaster (Working Man), which lived 1.9 to 1.5 million years ago, is another debated species in the *Homo* genus. Like all humans, it spent all of its time on the ground. It seems to be an intermediate form between *Homo habilis* and *Homo erectus*, so it depends on whether you are an anagenesist (lumper) or a cladogenesist (splitter) as to whether it deserves separate species status.[122]

There is no debate that *Homo erectus*, which lived from 1.8 million years ago until 50,000 years ago, is human. Its brain was larger, it used more advanced tools, it knew how to use fire for cooking, and it was clearly designed

An interesting genetic change occurred in the hominid lineage approximately 2.4 million years ago, which correlated with the enlargement of the human brain and may even have had a causal relationship. The myosin heavy chain gene (MYH16) mutated, resulting in a significantly smaller masseter (jaw) muscle. The anchoring of this huge muscle in chimps and other apes results in a large depression of the skull. Relieving this skull growth pressure may have enabled the subsequent enlargement of the brain and skull.

See: H. Stedman, B. Kozyak, A. Nelson, et al., "Myosin Gene Mutation Correlates with Anatomical Changes in the Human Lineage," Nature 428 (2004): 415, DOI 10:1038/nature02358.

for long-range walking and running. In fact, it walked to Europe and Asia.

We have now finally arrived at the *Homo* genus after about 3.8 billion years of life on earth, and approximately 2 million years ago. Although it seems like a momentous event in our history, it wasn't really an event at all. Rather, it was a slow, convoluted march through millions of years of evolutionary changes that led in many directions, many of which were dead-ends with other directions leading to the current great apes and other existing genuses. It is a pathway that is still shrouded with debate on the specifics. Clearly, what distinguishes *Homo* from pre-*Homo* is our advanced brain, the structural and metabolic changes that support the brain, and the behavioral and cultural consequences of it. These were all hugely relevant changes to defining who we are, and they certainly set the stage for what we will become. We now need to dig deeper into the finer details of the changes within the *Homo* genus to understand how *Homo erectus* and the other early humans became *Homo sapiens*. This will inform our conclusions regarding *Homo nouveau*.

Chapter 8

Out of Africa

"Until about 1.8 million years ago, all of the humans on earth were in Africa."

Until about 1.8 million years ago, all of the humans on earth were in Africa. For whatever reason—whether in search of better food sources, seeking freedom from conflict and predation from other species, or simply the curiosity and wanderlust of an imaginative brain—a small group of *Homo erectus* humans made their way into the Middle East and Western Asia. Whether that was by walking through the Sinai Desert or crossing what is now the Bab-el-Mandeb Strait between Djibouti and Yemen is still being debated. This momentous event is sometimes referred to as "Out of Africa 1." They lived in the Middle East and Western Asia for the next one and a half million years or so. They continued to evolve there and in Africa, and their descendants finally developed brains as large as today's *Homo sapiens*. Although the fossil record is somewhat unclear, at some point during this period, a new species, *Homo heidelbergensis* (named because the earliest fossil was discovered near Heidelberg, Germany) may have emerged from *Homo erectus* and interbred with them. The two species also continued to spread out in two directions, some going to Europe and some going to Eastern Asia.

The muddle in the middle

Homo heidelbergensis typifies the difficulty paleontologists have in distinguishing species in the *Homo* lineage. Its skull is intermediate in size between *Homo erectus* and modern *Homo sapiens*. There is evidence that its tools were slightly more advanced than *Homo erectus*, and they may have buried their dead, which would indicate a cultural change as well. None of these changes can be determined with certainty, and these distinctions (as in other distinctions in our lineage) are somewhat arbitrary and continue to be debated. Discussion of *Homo heidelbergensis* is often referred to as the "muddle in the middle."[123]

What is also a muddle is whether this group from "Out of Africa 1" ultimately evolved into our closest *Homo* relatives, the Neanderthals and Denisovans, or died out without further speciation. Genomic studies indicate that *Homo neanderthalensis, Homo denisova, and Homo sapiens* have a common ancestor. Our genomes overlap by 99.7 percent—certainly more than we overlap with today's chimps and bonobos. The common ancestor is not known with certainty. *Homo heidelbergensis, Homo erectus, Homo helmei,* and *Homo antecessor* (or simply archaic *Homo sapiens*) have all been postulated as the common ancestor based on various fossil evidence and semantic debates.

...meanwhile, back in Africa

We do know that some *Homo erectus* remained in Africa after "Out of Africa 1" and continued to evolve there until about seven hundred thousand years ago. Fossil evidence from various places in Africa indicate at least two phases of evolution occurred from about seven hundred thousand years ago until about two hundred thousand years ago, when anatomically modern *Homo sapiens* fossils have been found. These pre–*Homo sapiens* phases have been characterized by some as "early archaic *Homo sapiens*" and "late archaic *Homo sapiens.*" Others believe they represent speciation events from *Homo erectus* to *Homo heidelbergensis* to *Homo helmei*, and finally to *Homo sapiens* around two hundred thousand years ago.[124] This illustrates the problem with defining species, particularly in our lineage from very sparse fossil evidence. In any case, *Homo erectus* and its successors in Africa continued to develop a more complex and advanced brain capable of going far beyond simple tools and small-group hunting and gathering. They became dominant forces in their ecology. Certainly, by two hundred thousand years ago in Africa, our morphology, including our large brains, had evolved into modern *Homo sapiens.*

Possible immediate ancestor to *Homo sapiens*
Homo erectus
Homo heidelbergensis
Homo helmei
Homo antecessor
Homo sapiens idaltu
Archaic *Homo sapiens*
(see appendix 6)

Homo sapiens was an African birth...slowly

What does seem clear is that the emergence of *Homo sapiens* in Africa was not a sudden speciation event, but rather a more gradual evolution through a number of phases over hundreds of thousands of years. These did not necessarily all occur in one part of Africa, which is a huge continent that was sparsely populated by these humans. Whether this led to allopatric speciation events or simply a more continuous evolution of subspecies is debatable and somewhat arbitrary.

Most of our current literature dates the emergence of modern *Homo sapiens* to about 200,000 years ago based on fossils found in Ethiopia in eastern sub-Saharan Africa. However, a recent report of fossils found in a cave called Jebel Irhoud in Morocco (north-western Africa) may push that date back to 300,000 years ago.[125] This finding also suggests that our speciation event could have occurred in multiple places in Africa rather than in a single area. The Moroccan skulls are slightly more elongated (i.e. more "archaic") than the Ethiopian skulls raising some question as to whether they are truly modern *Homo sapiens* or some intermediate with our predecessor. This simply illustrates the difficulty in our ability to be precise about our origins in Africa both in timing and location.

Sally McBrearty and Alison Brooks point this out in their wonderful review of the origin of *Homo sapiens.*[126] To truly trace the speciation of *Homo sapiens*, as with any species, one must look at the preponderance of evidence both in the direct fossil record as well as the indirect record relating to species behaviors. Figure 15, which is taken from their paper, lists the numerous behaviors that are associated with the emergence of modern humans that are seen in the indirect fossil record. As can be seen, these behaviors originated at various times in Africa, spanning a period from about 280,000 to 50,000 years ago. There was no sudden or abrupt "human revolution," but rather a gradual transition from archaic to modern humans.

FIGURE 15

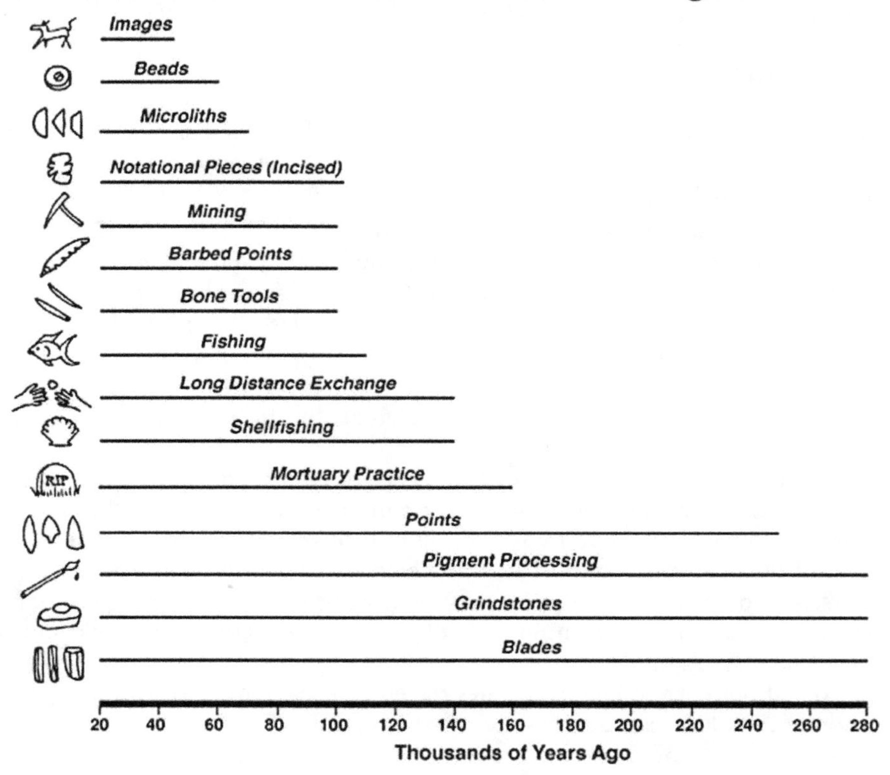

From: S. McBrearty, S. Brooks, "The Revolution That Wasn't: A New Interpretation of the Origin of Modern Human Behavior," *Journal of Human Evolution* 39 (2000): 453.

Neanderthals and Denisovans

Getting back to the question of the emergence of the Neanderthals and Denisovans in Europe and Asia, the alternative to the "Out of Africa 1" scenario is that there were subsequent "Out of Africa" events leading to their emergence. We do know that *Homo neanderthalensis* (Man from Neander Valley) branched away from the common ancestor to *Homo sapiens* perhaps as long ago as seven hundred thousand years; this is based on the extraction of DNA from a Neanderthal fossil found in a cave in Spain.[127] There is evidence that there was at least one other "Out of Africa" migration consisting of some of the archaic *Homo sapiens*. At the time of the split, which probably happened somewhere in the Middle East, the population of humans was calculated to be less than eleven thousand females.[128] The Neanderthals subsequently spread out from the Middle East to both Europe and Asia. They went as far north as

Great Britain (when there was still a land bridge with the continent, before the English Channel formed following the glacial melting). These hulking, large-brained humans dominated Europe for several hundred thousand years and finally became extinct about thirty thousand years ago.[129] They knew how to use tools and control fire for warmth, light, and cooking—skills carried over from *Homo erectus*.[130] They wore clothes and "almost certainly had spoken language."[131] They probably also back migrated to a small extent into Africa from the Middle East.[132]

In parallel with the emergence of the Neanderthals, another advanced species of humans emerged in Southern Siberia called the Denisovans (named after their discovery in the Denisova Cave in Siberia). We have very little fossil evidence of Denisovans, but we have been able to extract their DNA from those few fossils. This is the first time that a new hominid species has been declared solely on the basis of genetics.[133] They emerged about the same time as the Neanderthals and coexisted and interbred with them for about 150,000 years throughout Asia.

What exactly is different about Homo sapiens?

It is difficult to glean from the fossil evidence exactly what changed between our immediate predecessor and *Homo sapiens* in Africa that warranted declaring us a new species. That would be very helpful to know because a similar change may create *Homo nouveau*. The genome evidence is starting to nail down this question a little better. It is only in recent years that we have been able to perform complete genome construction not only of chimps and humans, but also of many other species of mammals, birds, and other organisms. Using massive computer power and statistical analyses, we can infer a lot of things by comparing the complete genomes of different species. Genetic mutations are random, but over long periods of time they occur at a fairly consistent rate. This rate can be used to date the time when two existing species had a common ancestor by counting the nucleotide differences in overlapping portions from their genomes. In addition, because natural selection is operating on those changes, some parts of the genome will show more changes than others because those changes conferring a beneficial effect will be selected to be passed on to future generations. These so-called accelerated regions tell us a lot about which changes were important in speciation.

Katherine Pollard at the University of California, Santa Cruz, did such a genome comparison.[134] She found that there is a very small part of the human genome that has undergone the most rapid changes since we split from the chimps. That region is called HAR1 (human accelerated region 1). HAR1 consists of just 118 nucleotides (out of three billion in our genome). By looking at these same nucleotides in the genomes of chickens and other species, she could determine that those same nucleotides were stable in the previous three hundred million years in those other species. So something happened in the 5.4 million years since we split from the chimps to cause the changes in these particular nucleotides to be selected. Those changes must be related to the emergence of *Homo sapiens*. It turns out that HAR1 is important in regulating the genes that cause the human brain to develop the convolutions in our neocortex. It is these convolutions that provide the enormous increase in surface area of our brain, enabling much more complex behavior. They are epigenomic changes.

There were some other HARs that Dr. Pollard examined as well. These are related to our linguistic ability and our increased manual dexterity—two other key differentiators of our species. Thus, it appears that *Homo sapiens* emerged because of a relatively small number of genetic mutations in a relatively short period of time.

As Dr. Pollard states, "In other words, you do not need to change very much of the genome to make a new species."

A different type of whole genome comparative analysis was done at the University of Edinburgh.[135] They compared the human genome to eleven other mammals, including chimpanzees. They found a gene regulator, called miR-941, that exists only in humans. It is the only one they found that is unique to humans of the species studied. This regulator is involved with genes related to longevity and cognition. Because chimps don't have miR-941, this gene regulator must have emerged since the split. It is yet another way that the information explosion now occurring because of our ability to analyze the genomes of many species will pinpoint exactly what makes *Homo sapiens* what we are.

So far, it looks like it is all in the epigenetics, not the genes that make us humans rather than chimps.

Out of Africa 2

Regardless of what were the exact changes leading to *Homo sapiens,* there is little doubt that *Homo sapiens* emerged in Africa long after its predecessors had left Africa during "Out of Africa 1." Like their ancestors, *Homo sapiens* were not satisfied to remain within the confines of Africa. At various times, small numbers of them migrated out into the Middle East. Until recently, it was thought that fully modern *Homo sapiens* did not migrate to the Middle East until 50,000–60,000 years ago.[136] However, there is some new evidence from fossil teeth found in Southern China that modern *Homo sapiens* arrived there between 80,000–120,000 years ago, long before they made their way into Europe. It is possible that those that migrated between 80,000 and 120,000 years ago died out and were replaced by the later migration 50,000 to 80,000 years ago. This later migration is sometimes referred to as "Out of Africa 2," although we now know that there were multiple migrations between Africa and the Middle East. By 120,000 years ago, Neanderthals had already populated Europe from earlier migrations. It is speculated that a combination of the Neanderthals in Europe and the colder weather there may have delayed incursion of modern *Homo sapiens* into Europe.[137] The exact number who did this later "Out of Africa" migration of *Homo sapiens* could have been as small as a few thousand or even a few hundred, according to some genetic studies.[138] From this inauspicious beginning, the world was changed.

This time, *Homo sapiens* did not stop in Europe and Asia. By fifteen thousand years ago, they eventually populated all of the Americas.[139] They most likely followed a land route across the Bering land bridge, and then along the Alaskan and Canadian Coast into Canada and the rest of the Americas.[140] Those that populated Europe became known as the Cro-Magnons, modern *Homo sapiens* who coexisted with the Neanderthals there for as much as 10,000 years. (The term "Cro-Magnon" simply denotes the early modern *Homo sapiens* found in Europe and are not different from any other modern *Homo sapiens* from that period.) Likewise, they spread to what is today Indonesia, New Zealand, and Australia at a time when the sea levels were lower and land bridges existed to these areas, or there were shorter boat rides than today. In the process, they left nothing untouched in their wake. They probably used their superior brains to speak in some kind of language to each other, control fire, domesticate plants and animals, and form societies that weren't always on the go. They also probably wore clothing.

When did *Homo sapiens* get dressed up?

Clothing does not survive the way tools, pottery, wall art, and other indirect evidence does, so it took our clever *Homo sapiens* of today to figure this one out. They used the genetics of body lice to do that! Human body lice live only in clothing, and by applying our genomic dating methods to today's human body lice, they determined that human body lice first evolved between 83,000 and 170,000 years ago. That would correspond to the time before "Out of Africa 2," when *Homo sapiens* was still in Africa but after "Out of Africa 1," when other humans were in Europe and Asia.[141] Thus we probably wore some kind of clothing before we left Africa as *Homo sapiens* and carried those clothes (with their lice) with us. We also suspect that the Neanderthals in Europe wore clothes.

When did we start talking?

We still haven't found the smoking gun in our genes to nail down exactly when language occurred, but some evidence suggests it was prior to when *Homo sapiens* split from its predecessor species so that *Homo sapiens,* Neanderthals, and Denisovans may all have had similar language development.[142] The best evidence for this is found in the FOXP2 gene. In humans, a defect in this regulatory gene prevents normal language development. FOXP2 in *Homo sapiens* differs from the chimp version but is identical to that of Denisovans and Neanderthals. Also, the anatomy of the pharynx (voice box) in chimps and other apes, as well as the hyoid bone (which anchors the tongue), is significantly different from humans. The human version is more adaptable to the delicate alterations in phonation required for speech.

Robert Berwick and Noam Chomsky, in their book *Why Only Us: Lan-*

guage and Evolution, argue that the FOXP2 evidence may be misleading.[143] They state that language, as represented by the many languages spoken around the world, is the incorrect focus. That is, "talking" is the wrong focus. They argue that the fundamental purpose of language is the internal mental representation of concepts that allows us to think—to plan, to reason, to create, and to perform all of our other higher cognitive functions. Evolution created this internal language and it is the same in all *Homo sapiens*—regardless of what language we speak. In this sense, there is just one human language. The external communication function—what we call languages—is a secondary function. The ability to create sounds (or signs) with our highly tuned phonations is sensory-motor activity that developed later in many different variations, leading to the many different languages we have today. The regulatory gene FOXP2 relates to these sensory-motor functions and therefore is not related to the fundamental evolutionary change leading to language. In their view, this more fundamental evolutionary change occurred around eighty thousand years ago in Africa, long after the earlier migrations of *Homo* species out of Africa but before the final migration of modern *Homo sapiens.*

Homo sapiens dominate the Homo world

Homo sapiens do dominate the *Homo* world as evidenced by the fact that we are the only one left standing. It was during the later "Out of Africa" period beginning 50,000–120,000 years ago that modern humans went from "insignificant animals with no more impact on their environment than gorillas, fireflies or jellyfish"[144] to the dominant species in their ecology. It was then that *Homo sapiens* accelerated their differentiation in language and other cognitive functions from their fellow humans. All other human species became extinct by about 30,000 years ago.

How did this happen? Were the *Homo sapiens* some type of marauding group, spreading out and killing everything in their path? After all, the Neanderthals had big brains too—and bigger bodies. Neanderthals used fire and tools, were good hunters, cared for their sick and wounded, were well adapted to the colder environment, and had their own form of society.[145] In fact, they had survived previous cold climate change events before and were better adapted to cold weather than *Homo sapiens*. How did we out-compete them?

Our overlap with the Neanderthals in Europe lasted for 2,000–10,000 years. Although that sounds like a long time, in evolutionary terms it is a very short time. We don't have all the answers, but thanks to some of the most recent advanced *Homo sapiens* tools, we now have a lot more answers than we did a decade ago. Those tools include the ability to extract small amounts of preserved DNA from fossils of extinct species, replicate these small amounts into large amounts suitable for analysis using PCR (see chapter 12), and then apply rapid DNA sequencing and computer modeling to discover facts about our history that were impossible to discover only a few years ago.* This has

*Even more recently, tools have been developed to detect ancient DNA from sediments even in the absence of fossils. (See Science 356, no. 6338, (2017): 605)

opened up the new field of paleogenomics. And what better way to start this new field than to discover that we could get some DNA from fossils of two of our closest human relatives, Neanderthals and Denisovans? We now have complete nuclear and mitochondrial genome sequences from a few individuals of each of them. Because we can compare their genomes to larger samples of existing *Homo sapiens* from all over the world, we have learned amazing facts.

First of all, the DNA from current Europeans contains 1–2 percent of their DNA from Neanderthals.[146] A similar amount of Neanderthal DNA is contained in today's Asians. However, no Neanderthal DNA is contained in today's Africans.[147] Wow! This genome analysis confirms that there was a small amount of interbreeding between Neanderthals and *Homo sapiens* that must have occurred outside of Africa. Neanderthal DNA shows up in Europeans and Asians about the same amount, which suggests that the interbreeding could have occurred in the Middle East before *Homo sapiens* spread in the two directions. Similarly, some of today's Asians, particularly those in Oceania, contain as much as 3–6 percent of their DNA from Denisovans, whereas Europeans contain no Denisovan DNA. Native Americans do have Denisovan DNA. This pattern would confirm that *Homo sapiens* interbred with Denisovans to a small extent while they were on their way to North American through Asia.

Figure 16 shows one possible sequence of events in the timeline of evolution of the *Homo* genus. In this version, the Neanderthals and Denisovans emerged from an early migration of *Homo erectus* from Africa.

FIGURE 16

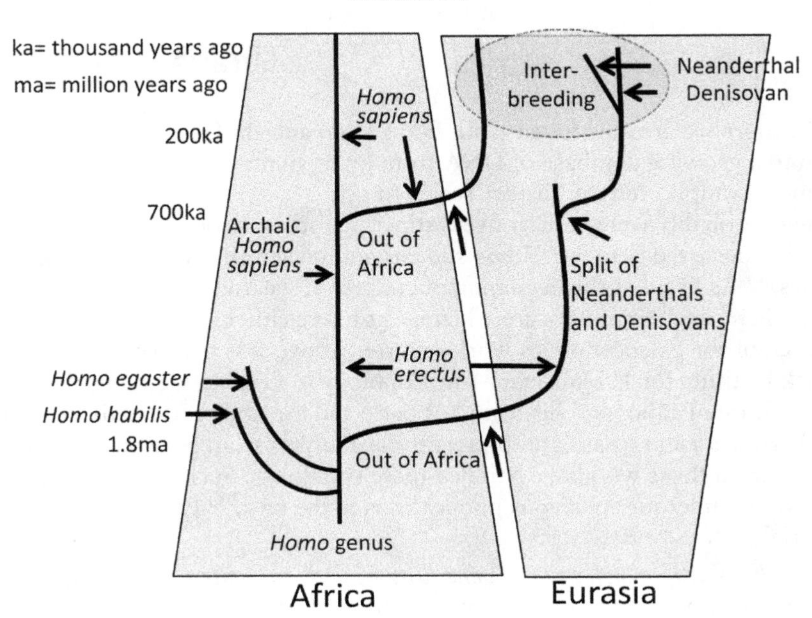

Figure 17 illustrates an alternative interpretation of the fossil and genomic evidence. In this version, the Neanderthals and Denisovans emerged following a later migration of archaic *Homo sapiens* from Africa. There are many other possible versions as well. Chris Stringer, in his recent review of the origins of *Homo sapiens*, illustrates three other variations on this theme.[148]

FIGURE 17

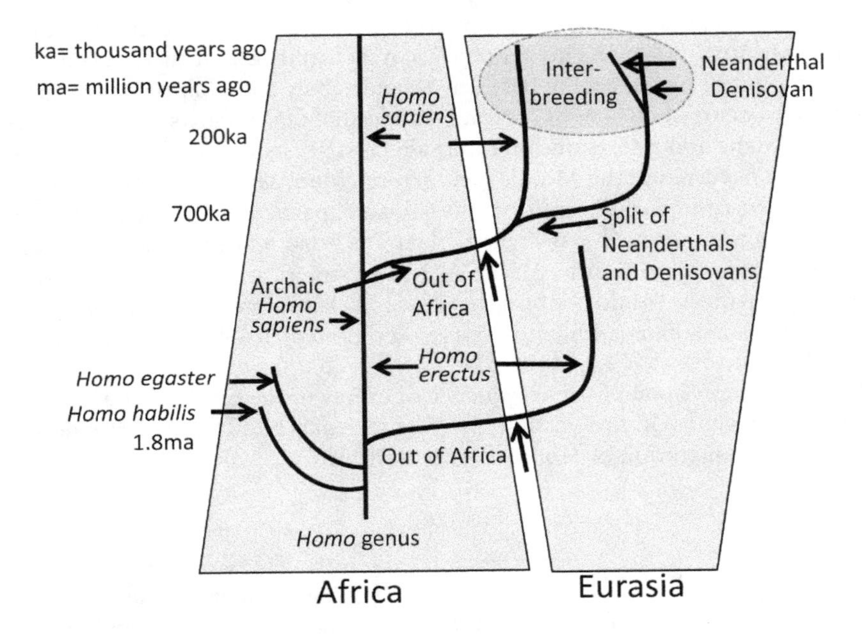

The more we are able to tease out DNA from ancient fossils and compare them to a growing database of DNA from living humans around the world, the more complicated our history seems to get.

There probably were at least five distinct periods and locations where interbreeding occurred between *Homo sapiens* and ancient (now extinct) human species.[149] As I said at the beginning of chapter 6, we don't know for sure how we got here, and we're not sure when we got here either.

Much of the paleogenomics work reported above was performed at the Max Planck Institute for Evolutionary Anthropology in Leipzig, Germany. Its director, Svante Pääbo, and his team has perfected the ability to reliably extract DNA from ancient fossils and protect these analyses from possible contamination from those who have handled these specimens. Such contamination has led to numerous erroneous publications in the past.[150] Pääbo said in an editorial in *Nature Reviews*,

> *The multiple instances of gene flow now documented among hominid groups show that modern humans were part of what one could term a "hominid metapopulation"—that is, a web of different hominid populations, including Neanderthals, Denisovans and other groups, who were*

linked by limited, but intermittent or even persistent, gene flow ... As the ancestors of modern humans were presumably part of the hominid metapopulation, one can imagine that the multiple genetic variants that set modern humans apart from other hominids originated not in one single African population but in many different African populations and, in rare cases, perhaps even outside Africa. Gene flow in the hominid metapopulation could then have allowed these genetic variants to come together in the explosive constellation that caused the modern human population to expand and replace everyone else.[151]

Pääbo's conclusions reinforce those of many others: "It is clear that the genetic 'recipe' for making a modern human is not very long."[152]

What's the point?

There are two takeaways from all of this historical discussion so far. The first is that it is very difficult to reconstruct when and where new species in our lineage have appeared. I do not believe we will have this problem going forward because unlike our predecessors, we understand speciation and genomics and will be witnessing our evolution in real time. Second, as I stated before, it takes very little change in our genomics to create a new species. That will certainly inform the answers in chapter 14.

For ten thousand years or longer, *Homo sapiens* coexisted with at least two other human species outside of Africa: *Homo neanderthalensis* and *Homo denisova*. What was that coexistence like? That part is speculation, but it may be a clue to how *Homo sapiens* will coexist with *Homo nouveau*. We know they interbred, but not much. Probably they mostly lived separately. As *Homo sapiens* moved from Africa to the Middle East, then Southern China, and finally Europe, they obviously encountered the other humans. It was at these border points or hybrid zones that occasional interbreeding occurred.[153] There is fossil evidence of hybrid individuals between *Homo sapiens* and Neanderthals in Southern Europe.[154][155] Was this normal mating? Casual encounters like today's hookups? Rape? We don't know the answer. One of the interesting DNA facts is that all of the Neanderthal DNA found in today's *Homo sapiens* genome is in the nuclear DNA—but there's none in the mitochondrial DNA. Does that mean that the only interbreeding was between male Neanderthals and female *Homo sapiens*? Did they rape our women? There has been at least one hybrid Neanderthal/*sapiens* fossil found where the mitochondrial DNA was Neanderthal, so that would suggest the opposite. The lack of Neanderthal mitochondrial DNA in the *Homo sapiens* genome of today may have more to do with the survival of the hybrid offspring and how they were integrated into the societies of their parents. The fact is that we don't have a large enough sample size of the extinct species to make definitive conclusions at this point. We know that we did not form a hybrid species with either the Neanderthals or Denisovans but continued to evolve separately. We survived; they didn't.

What was the barrier to interbreeding?

We know we interbred with Neanderthals and Denisovans, yet we evolved separately from them. What was the barrier that caused this near, at least incomplete, reproductive isolation? The different human populations evolved mostly apart from each other geographically, so one could call this allopatric speciation. On the other hand, there were these hybrid zones. Again it's speculation, but at least two sympatric speciation mechanisms probably occurred in these hybrid zones. First, the offspring of the hybrids, although viable, did not reproduce abundantly enough to allow complete hybridization of the two species. Perhaps they were less fertile (a "postzygotic" explanation). Another possible reason for the offspring not producing abundantly could also account for the low incidence of interbreeding in the first place. The two species simply might not have been attractive to each other (a "prezygotic" explanation).[156] After all, they looked different. *Homo sapiens* were smarter, had darker skins, and they may even have smelled differently. As the Neanderthal experts Dimitra Papagianni and Michael Morse say in their book *The Neanderthals Rediscovered*, "There is no getting past the fact that to a modern human, Neanderthals probably looked ugly."[157] We probably looked ugly to them also.

Maybe the hybrid children were ostracized or sacrificed. Were hybrid offspring of Neanderthal mothers integrated differently into their society than hybrid offspring of *Homo sapiens* mothers? Were the mothers even aware of the paternity if they had mated with males from each species? Neanderthals and Denisovans may have had some kind of language, but as discussed earlier, that is uncertain given our current state of knowledge concerning the evolution of language. Even if they had language, it surely would have been different from *Homo sapiens* language. There is still a lot to learn, and as we get more evidence from the extinct genomes, some of this may become clearer. Figure 18 is a rendering by Nicolas Primola of a Neanderthal man compared to a modern human man.*

---

*Licensed through Shutterstock.com, 6/20/16, image #310746203.

Figure 18

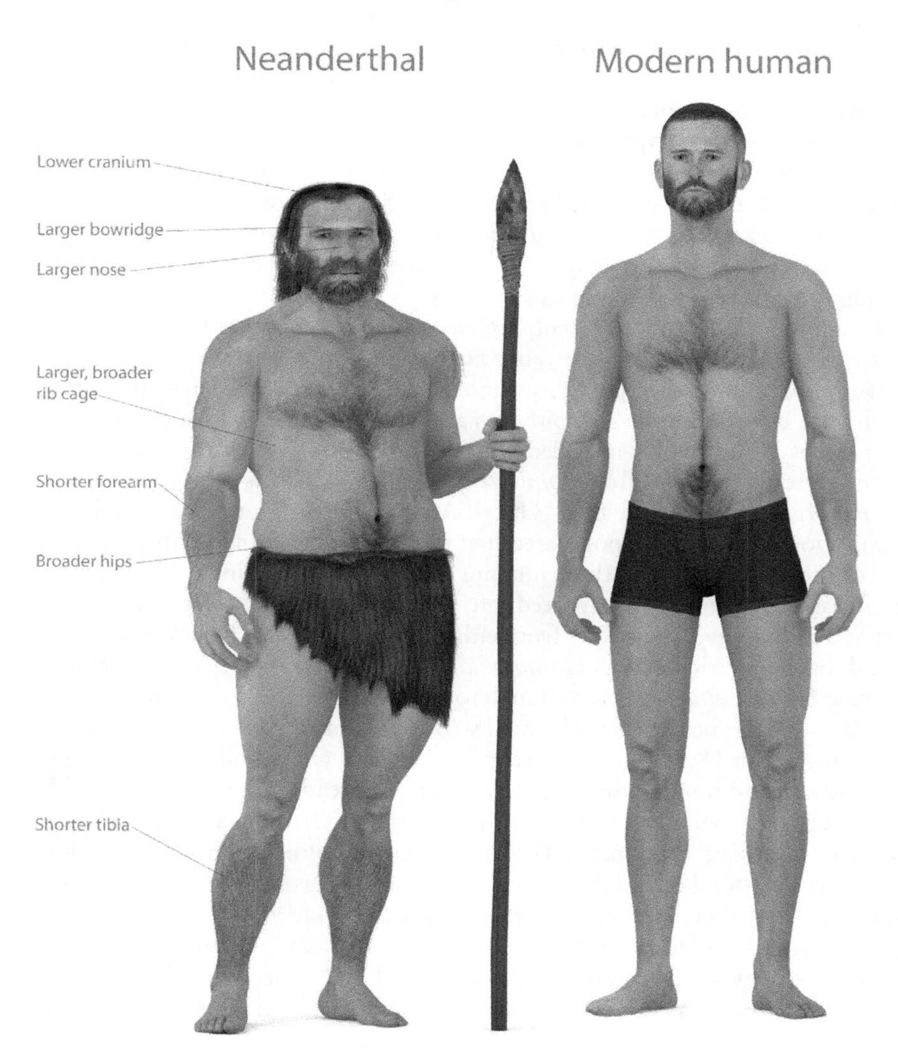
Neanderthal
Modern human
Lower cranium
Larger bowridge
Larger nose
Larger, broader rib cage
Shorter forearm
Broader hips
Shorter tibia

There is some speculation and even a little evidence that the small amount of interbreeding had some beneficial effect on *Homo sapiens*. Remember that the Neanderthals and Denisovans had already inhabited the Middle East, Europe, and Asia for thousands of years before *Homo sapiens* arrived. This allowed them to acclimate to their colder local ecologies, develop immunities to local pathogens, adapt digestion to local plants, and adapt in other ways to the environment. Some of their genes could have been helpful. There is some evidence that some of the beneficial Neanderthal genes related to immunity have been passed on to *Homo sapiens*.[158] Neanderthal genes are also thought to make us more susceptible to type 2 diabetes,[159] as well as depression and certain precancerous skin diseases.[160]

In summary, our path to *Homo sapiens* went from a chimp-like creature five to seven million years ago to our current modern form dating back to about two hundred thousand years ago. The first several million years of that path took us through the Ardipithecus and Australopithecus phases to the first *Homo* genus, which included *Homo habilis, Homo ergaster,* and then *Homo erectus* (and possibly *Homo heidelbergensis*). It was a small group of *Homo erectus* that led the first "Out of Africa" wave about 1.5–2.0 million years ago, when humans populated first the Middle East, then Europe, and Asia. In Europe and Asia, these humans may have evolved into Neanderthals, and in Asia they may have evolved into the Denisovans. However, *Homo sapiens* did not evolve directly from either Neanderthals or Denisovans. Instead, back in Africa, it was *Homo heidelbergensis, Homo Helmei,* or simply archaic *Homo sapiens* that continued to evolve from *Homo erectus* into our modern-day *Homo sapiens*. The final wave of "Out of Africa" happened probably about 50,000-120,000 years ago when a very small number of our very advanced-brain, clothed, modern humans spread throughout the world, interbred to a small degree with Neanderthals and Denisovans, and eventually replaced all the other humans. The most recent genomic studies seem to indicate that almost all of today's humans derive from this final wave of migration and that the descendants of the earlier waves died out.[161] [162]

Although I have described two major "Out of Africa" waves, there likely were other waves between the two as well as after.[163] There were likely some "Into Africa" waves as well.[164] [165] For now, figures 16 and 17 must be considered just two of numerous possibilities about our exact ancestry rather than proven fact. The only absolute fact is that all of our immediate ancestors and nearest *Homo* relatives are extinct, and we aren't.

How did Homo sapiens dominate?

I asked the question earlier: How did we out-compete these other humans? We know *Homo sapiens* today is capable of ethnic cleansing. Maybe that's what happened; at least one article speculates that this was the case.[166] But why didn't they ethnic cleanse us? Neanderthals were bigger and stronger, had bigger brains, and were already adapted to the colder northern climate. We don't know the answer. One theory attributes our success to our advanced development of language between thirty thousand and seventy thousand years ago. This enabled *Homo sapiens* to cooperate and plan to a far greater extent than any other human species. It was the so-called cognitive revolution;[167] more on this in chapter 11. We also probably had better weapons. There is evidence that at this time, *Homo sapiens* was the only species that used advanced projectile weapons like spear-throwers (atlatls) and bows and arrows for hunting.[168] Killing at a distance overcomes the disadvantages of smaller body size and strength. Neanderthals did use less advanced projectile weapons like spears, which date back at least four hundred thousand years, before modern *Homo sapiens* existed. There is no substantiated evidence that the advanced projectile weapons of *Homo sapiens* were used against Neanderthals directly, but they certainly would have made *Homo sapiens* better competitors in hunting prey. Another theory regarding Neanderthal extinction is that *Homo sapiens* brought diseases from Africa into Eurasia to which the Neanderthals lacked immunity. The resulting potential epidemics could have weakened them.[169]

Pat Shipman, an anthropology professor at Pennsylvania State University, discusses the extinction of the Neanderthals at length in her book *The Invaders*.[170] During the period preceding the extinction of Neanderthals, there was a cooling climate change that almost certainly stressed the Neanderthals by limiting their normal prey and their natural environmental flora. Some attribute their extinction entirely to this. However, Ms. Shipman argues that the Neanderthals had survived similar cooling periods, and although they were stressed and weakened, it took the invasion of *Homo sapiens* finish them off.

She outlines the normal sequelae when an invasive predator species enters a new ecological area, and she argues that *Homo sapiens* had the same impact as any other invasive predator species. That includes direct competition for the same prey and space as other predators, which in this case would have included the Neanderthals. An illustrative example she cites was the reintroduction of wolves into Yellowstone National Park and the negative impact that it had on coyotes. In the case of *Homo sapiens* invading the territory of the Neanderthals, our diets were the same, meaning that we went after the same prey, particularly the larger animals (including mammoths). In essence, she argues, *Homo sapiens* garnered the market on mammoths.

Ms. Shipman cites three major advantages that *Homo sapiens* had over the Neanderthals. First, *Homo sapiens* were the only hominids that used bone needles to sew clothing with cords made from various fibers. Their more efficient protection against the cold was a clear enabler of all activities. Second, their mastering of advanced projectile weaponry made them superior hunters. Finally, and to Ms. Shipman the *coup de grace*, *Homo sapiens* domesticated wolves into doglike companions that were superb collaborators in tracking and cornering large prey like mammoths, allowing for easier kills.* These

*Actually, some believe that wolves domesticated themselves. That is, wolves, which are natural scavengers, learned that befriending rather than threatening humans led to much greater access to food. Thus certain ones became partners with humans naturally. (See R. Coppinger in The Domestic Dog, J. Serpell, Ed., Cambridge University Press, 1995)

animals could also help guard the kills from other predators and scavengers. These three cultural, brain-related advantages made the difference between extinction and major expansion. The extinction occurred in many other predator species, including the Neanderthals and cave bears.

Left out of Ms. Shipman's story is the known fact that *Homo sapiens* and Neanderthals interbred to a small extent. Given that we were an invasive species, outperforming the Neanderthals in finding food, clearly threatening them, and forcing them into smaller and smaller enclaves, it is unclear what the interbreeding was all about. The story is unfinished. But it is clear that something different was happening to the *Homo sapiens* brain from about fifty thousand years ago onward that did not happen to the Neanderthal brain—at least not to the same degree as happened to *Homo sapiens*. We became much more capable of symbolic thought, as evidenced by more elaborate wall paintings, jewelry, musical instruments, and other indirect evidence. Our childhoods were longer (i.e., *Homo sapiens* matured more slowly, which is thought to enable better cognitive development). We became more creative in tool development. Our language was probably much more advanced. This more advanced cognitive capability allowed better foresight and planning, which ultimately led to our survival.[171] At least that's the theory. It is still a raging debate as to how cognitively advanced the Neanderthals were. Recent evidence seems to prove they made jewelry—one of the indicators of advanced cognition.[172] Neanderthal intelligence is being judged through the retrospective biases of *Homo sapiens*. We may never know for sure.

Some conclusions

It is a curious fact that the path to *Homo sapiens* since our chimp-like ancestor has left every species extinct from which our line has evolved. Is that foretelling the fate of *Homo sapiens*? After *Homo nouveau* emerges, will *Homo sapiens* be another dead-end? We'd have to survive another million years or so to match the record of *Homo erectus*' longevity on earth. How likely is that?

The bigger question regarding the answers is what we have learned about the speciation events leading up to *Homo sapiens*, and how likely they are to repeat themselves. The process certainly involved the usual combination of random genetic changes operated on by natural selection, which have operated since the beginning of life on earth. As will be seen in chapter 11, those forces are surely continuing today. But speciation requires both genetic change and the introduction of barriers to interbreeding.

With regard to the genetic changes that occurred between our immediate ancestors and us, we don't have sufficient DNA for the species from which we directly evolved: *Homo heidelbergensis, Homo erectus*, or archaic *Homo sapiens*. But we do have the DNA from the Neanderthals and Denisovans, which tells us a lot. What that tells us, as cited earlier, is that it doesn't take much genetic change to make a new species. Only 0.3 percent of our genome differs from the Neanderthals. We also have the HAR and other genomic analyses of living species that points to a few key changes related to brain development, language, and manual dexterity.

Even when we compare our genome to our closest living relatives, the chimps and bonobos, it is incredible how close our genomes are, especially in the protein-coding portion of our genomes (classically called genes). The big differences are in the epigenomic portions. According to the Ensembl genome database, a search in September 2015 showed 22,719 protein-coding genes for *Homo sapiens* compared to a similar number, 18,759 for chimps. But there were 26,781 noncoding genes in the humans compared to only 8,668 for the chimps. Some of that difference is because there is much more research being done on the human genome than chimps, and all the chimp genes are not yet identified and posted on Ensembl. Still, the difference between the coding genes and epigenome is striking.

There is much more genomic research being done today on mice than on chimpanzees because mice are more plentiful, inexpensive, and easier to study. Also, there isn't as much social objection to mice research as there is to chimpanzee research. A comparison of the human Ensembl genome database to mice is even more revealing. Mice have 22,671 protein-coding genes reported—almost identical to the number in humans—but still only 12,692 noncoding genes. It is clear what makes humans human is primarily in the epigenome. I will look closely there for the *Homo nouveau* answer.

What is also clear from our history is that most of the differences between us and our immediate predecessors are cultural: creation of more sophisticated tools (including weapons), formation of larger communities, domestication of plants and animals, use of art and language, and societal traditions and organizations.

All of this reflects changes in our brains and our ability to support a complex and high energy-consuming brain. I will also look closely there for the answers.

In terms of reproductive barriers, the second ingredient necessary for speciation, that becomes more problematic for *Homo sapiens*. All of our predecessor species, going back to primates, had very small numbers compared to today's humans. All of our lineage occurred in Africa, a very large continent. Clearly, allopatric speciation was a major factor in the development of speciation in our lineage. Pockets of ancient hominids could freely diverge genetically because of little or no contact between the separated groups, and thus there was no intermingling of the genetic pool for many generations—enough to diverge in ways that became barriers to interbreeding. We do not have that luxury today. We do know that there were some significant contact points and interbreeding, such as with the Neanderthals and Denisovans. What prevented more complete mixing of the gene pools and a breakdown of the reproductive barriers is not known. As discussed earlier, in the case of the Neanderthals, the transition was more likely that of an invading species edging out the existing endemic species through direct competition for prey and space. Keep in mind, however, that neither Neanderthals nor Denisovans were direct ancestors to

Homo sapiens. Our cladogenesis event occurred in Africa from either *Homo erectus*, *Homo heidelbergensis,* or perhaps some other closely related predecessor. Because we have little genomic information of these predecessors, we don't know the extent (if any) of interbreeding, and neither have we any information on their barriers to interbreeding. I will speculate on what barrier or barriers there will be between *Homo nouveau* and *Homo sapiens* in chapter 14.

What is clear is that the emergence of *Homo sapiens* was not the momentous event some people like to think it was. Robert Foley said it well at the end of his book *Humans before Humanity*:[173]

> *As we have seen here, human evolution is no blinding flash and no special creation. Man did not make himself, nor woman herself. Both are the product of countless events in the daily lives of the hominids. There is no magic ingredient in human evolution, and no substitute for knowing the details of what happened – where and when and why. Small, insignificant earthquakes in Africa, or particular demographic trends in Europe, are responsible for what happened in evolution. We should not let the uniqueness of our species dupe us into believing that we are the product of special forces. Cosmologists studying the origins of the universe need to think in terms of a big bang. Evolutionary biologists are better off with a bout of hiccups. If we had been privileged enough to observe the origins of our species and our lineage, we would have been struck by one thing – nothing very much happened.*

Part 2

Getting to *Homo Nouveau*

Chapter 9
How Might *Homo Sapiens* Evolve?

"It has taken us 200,000 years to get to this point...Now I am leaving the realm of facts and history and heading into speculation."

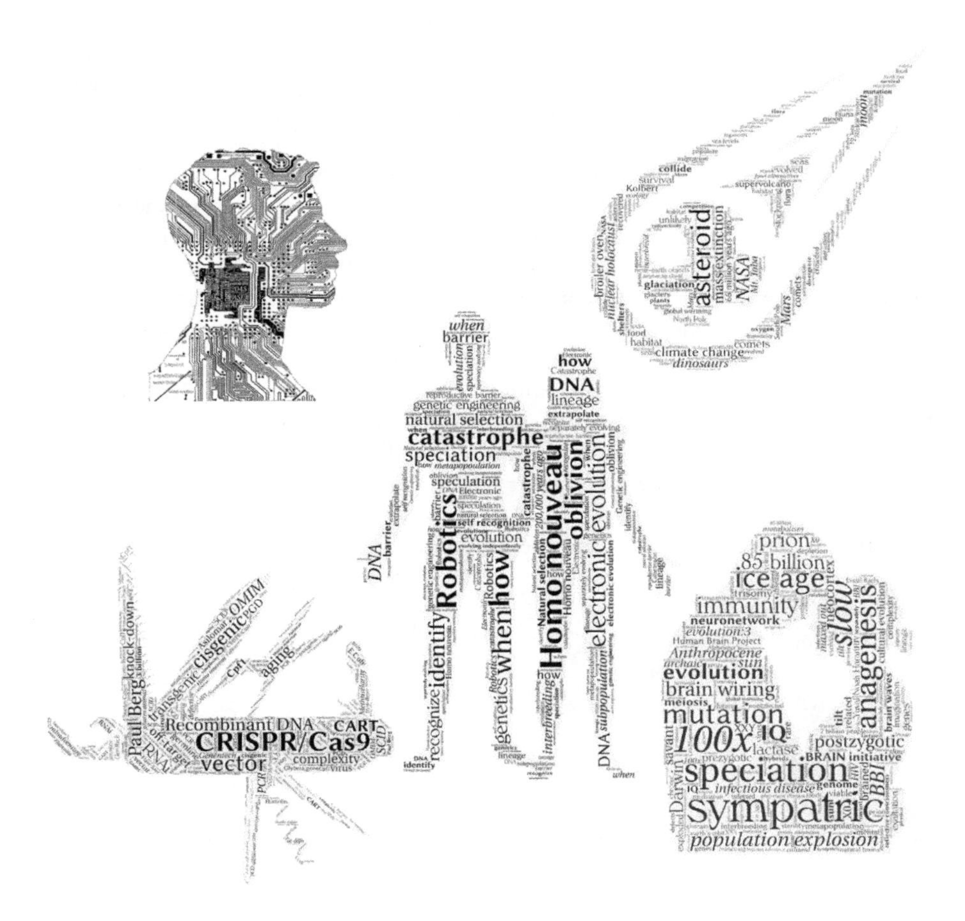

The basic concepts of species, speciation, genetics, evolution, and natural selection have now been discussed. I have reviewed that knowledge as it applies to *Homo sapiens*. I am now ready to think about how *Homo nouveau* might arise. I have the definition of species that I will use: a segment of a separately evolving metapopulation lineage. This means that I will need to be able to clearly identify a subpopulation of us that is evolving independently of the rest us, with some barrier to interbreeding. This also implies that I can identify the barrier or barriers to interbreeding that allowed this speciation to take place. Finally, I will need to have this process proceed long enough to insure it is a true lineage, not just a one-shot deal or some single point in time.

As was seen from chapters 6–8, the process of going from a chimp-like ancestor to *Homo sapiens* took millions of years and dozens or more side trips to oblivion. We are not talking about a repeat of that entire process to achieve the answers. Instead, it is probably more like the process that occurred in Africa about 200,000 years ago, when *Homo sapiens* emerged from a previous *Homo* species, although we're not exactly sure what that species was. Whatever that species was, it didn't understand the concept of species, speciation, barriers to interbreeding, metapopulations, genetics, DNA, or anything else related to this process. We do. It has taken us 200,000 years to get to this point, but we will be able to make the call when *Homo nouveau* has arrived. If and when that happens, we will be the first species to be able to recognize such an event.

At this point, I am leaving the realm of facts and history as best I can know it and am heading into speculation. However, it will be speculation based on this history. The intent is to create not a science fiction story; the intent is to speculate based on science. I will be trying to extrapolate from what we know, not deviate from it.

I will divide the discussion into the following broad mechanisms for possible speciation of *Homo nouveau*.

1. Catastrophe driven
2. Natural selection driven
3. *Homo sapiens* driven
   a. Genetic engineering
   b. Electronic

Chapter 10

Catastrophe

"After every mass extinction (and there have been five), many species that survived recovered in population and evolved. After millions of years, there were actually more total species on earth than before the mass extinction. In fact, we are one of them."

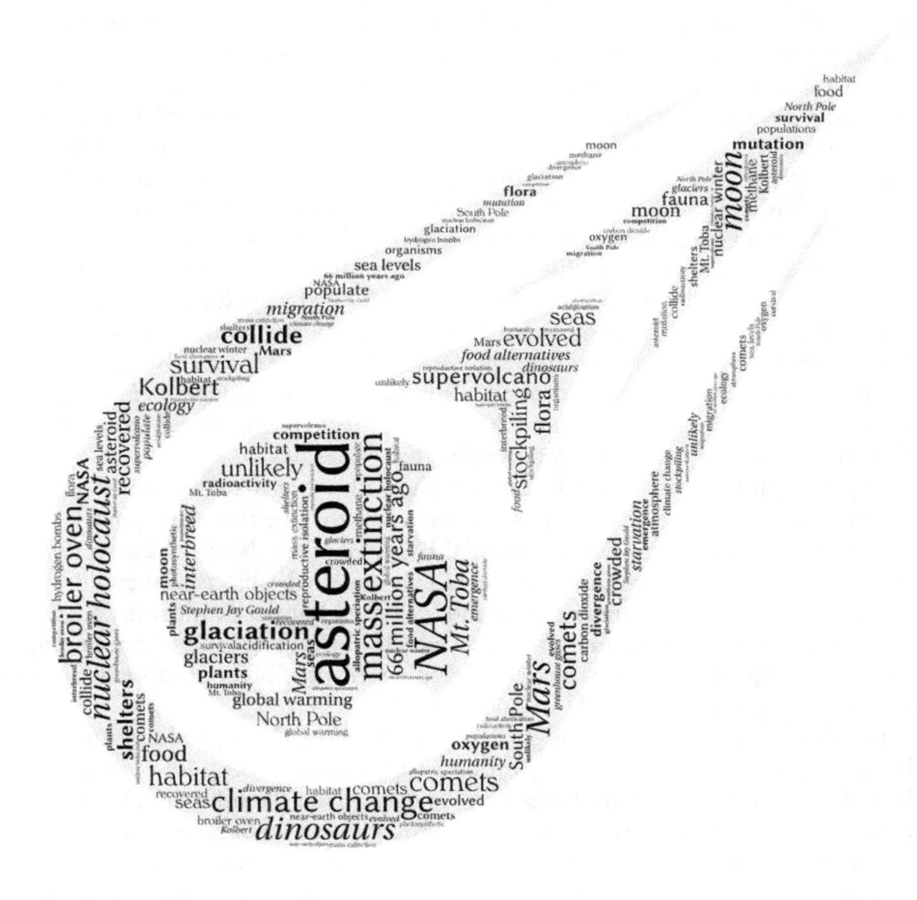

Most people are aware that there was a catastrophe about sixty-six million years ago that wiped out the dinosaurs. A large solid object collided with earth (called a bolide—technically not an asteroid as some people refer to it), causing massive extinctions. It not only killed the dinosaurs but about three-fourths of all living animal species at the time, particularly those larger than a cat.[174] If that happened again today and it wiped out all *Homo sapiens*, that would be the end of our story, and the questions of this book would become moot. There would be no *Homo nouveau.*

One could argue that that is not necessarily true. After every mass extinction (and there have been five), many species that survived recovered in population and evolved. After millions of years, there were actually more total species on earth than before the mass extinction. In fact, we are one of them. The question is whether any of the species that were wiped out has reemerged?

We know of no instance where one of the millions of species wiped out sixty-six million years ago has reemerged from total extinction. Even more to the point, given the randomness of evolutionary change, the likelihood that evolution after a catastrophe would repeat itself in a manner to produce anything close to *Homo sapiens* seems extremely remote. We were pretty unlikely in the first place. As the late Stephen Jay Gould said in his book *Full House*, "Perhaps we are, whatever our glories and accomplishments, a momentary cosmic accident that would never arise again if the tree of life could be replanted from seed and regrown under similar circumstances." He goes on to say that if we could ever replay the game of life again, "the vast majority of replays would never produce (on the finite scale of a planet's lifetime) a creature with self consciousness."[175] If *Homo sapiens* gets wiped out by any kind of catastrophe, life on earth would go on, and new species would appear (probably in greater quantity than before the catastrophe), but it won't be anything like *Homo sapiens*. The whole point of this book about our recognizing *Homo nouveau* becomes moot in the face of a catastrophe that wipes us out.

There might be a different post-catastrophe path other than extinction of *Homo sapiens* that is relevant. A bolide event like the one that wiped out the dinosaurs and most other species could happen again. NASA is currently tracking more than fifteen thousand so-called near-earth objects that could cause various levels of damage, up to and including human extinction, should they collide with earth. So far, none is on a collision course, but there are many more such objects not yet discovered, including the possibility of more difficult-to-track comets that could be out there.[176]

Elizabeth Kolbert, in her book *The Sixth Extinction*, has argued that the bolide impact was not a natural selection event; rather, it was pure random luck that determined which species lived or died.[177] On the other hand, one could argue just the opposite. Clearly something about an organism allowed it to survive, and that something may well have been related to its genetic makeup. I don't see that event as qualitatively different from a glaciation event or other major climactic change with regard to its natural selection capability. All evolution, in a sense, is pure random luck.

What if the catastrophe is not as complete?

Two questions arise. Could some *Homo sapiens* survive a future bolide impact that wipes out most other species? And could that lead to *Homo nouveau*?

The most obvious answer to both questions is no. If indeed the entire earth becomes a broiler oven, as may have happened last time,[178] there is no reason to believe any of us would be spared. But what if the bolide is smaller, and its effect is not uniform throughout the world? The one advantage we have over all the dinosaurs and other larger animals that existed sixty-six million years ago is that we are spread out over the entire globe in virtually every land-based habitat that exists. Conceivably, there could be a lucky outpost or two that doesn't suffer the full impact of the initial collision.

The next question is whether we could then survive the ensuing dark winter, when the food supply dwindles. Again, that seems unlikely. However, we are *Homo sapiens*, not dinosaurs. We might be clever enough to figure something out. The one way that I could imagine that happening is that our advanced technology a thousand or more years from now, when the next big one hits, would be able to predict an impact event long in advance. That'd be long enough to prepare ahead of time for the impact by building underground (or better yet, underwater) shelters and stockpiling the necessary food and other supplies. Or maybe we would even build an outpost on the moon for sufficient numbers of us to temporarily sit out the disaster until things became livable again on earth ("Out of Earth 1"). Of course, that's assuming that our advanced technology couldn't intercept the bolide and divert it before impact.

One of the most comprehensive and serious investigations regarding human survival following a dark-winter-type catastrophe is provided by David Denkenberger and Joshua M. Pearce in their book *Feeding Everyone No Matter What: Managing Food Security After Global Catastrophe.*[179] Although they point out that it is unlikely that a preplanned food storage solution would be sufficient for human survival, they outline several food production alternatives that would work, even in the absence of sunlight on the order of magnitude that would occur following a dark winter scenario.

The scenarios they examined included bolide impact, supervolcano, and nuclear holocaust. The food production methods include various techniques to extract calories from available resources that remain following the catastrophe. These include utilizing certain kinds of bacteria, techniques to extract food from dead plant material, growing mushrooms, cultivating certain animals for food that can digest cellulose from wood and other organic material, and various other processes. Their conclusion is that our clever *Homo sapiens* would figure it out.

So assuming that some of us do manage to survive, the next question is whether such a catastrophe in some way leads to *Homo nouveau*. It seems unlikely. Consider what is known about speciation as discussed in chapter 4. What if there were two physically separated surviving groups of *Homo sapiens*—say, one at the North Pole and one at the South Pole, or one on the moon. If they stayed separated long enough, even thousands of years (which would be on the short side for speciation), the two groups could evolve separately and ultimately evolve into two separate species by allopatric speciation. Which one would be called *Homo sapiens* and which would be called *Homo nouveau* is an interesting question, but surely they would both be intelligent enough to recognize what had happened.

This would be consistent with previous mass extinctions. In the subsequent millions of years, there was a rebound in the number of species evolving from those that survived the mass extinction. It was as though a void was created that allowed new species to emerge from the surviving ones and fill the void more rapidly than in previous periods.[180] This could have been because of the low numbers of individuals allowing mutations to have a larger relative effect, lack of competition, and greater space available to allow migration to separate locations, allowing allopatric speciation. Some believe that mammals and eventually humans would not have risen to our current prominence had it not been for the mass extinction event sixty-six million years ago.[181]

I don't envision such a scenario with *Homo sapiens* in a future similar disaster. In the case of two or more separate colonies of *Homo sapiens* following such a disaster, I see no reason why they would remain separate. Surely all of our technological know-how would still be preserved in the cloud servers on satellites. We could still build land vehicles, boats, and even aircraft; in fact, our thoughtful planning ahead would have stockpiled them in the shelters. There would be no reason for the two or more groups to not meet and interbreed over the ensuing years—long before allopatric speciation would have time to create a barrier. It is true that the greatly reduced population size of any surviving groups would make it more likely that a set of mutations causing divergence would populate a given group, but still, the time frames don't seem likely to be long enough. I'm having a difficult time making this one work. I think there are better candidate mechanisms for getting to *Homo nouveau.*

What about other types of catastrophes?

So far in this chapter, we've only talked about a catastrophe involving an extraterrestrial collision event. There are other types of catastrophes, of course. A supervolcano, such as the one that erupted seventy-four thousand years ago on Mount Toba in Indonesia, could have a similar effect. The Mount Toba eruption may have wiped out most of humanity.[182] What about a nuclear holocaust? It would also have a lot of similarities to a collision event in that it would be quick, involve immediate loss of life, and be followed by a longer period of a nuclear winter in which the skies would be darkened. It would have to be pretty big to equal the magnitude of the bolide collision sixty-six million years ago, but we have one thousand hydrogen bombs stockpiled between the United States and Russia. Detonation of this number of hydrogen bombs would approximately equal the bolide impact.

One significant difference would be the presence of radioactivity, which would increase the mutation rate. Perhaps, as in the scenario discussed with the bolide collision, if there happened to be two isolated surviving groups of *Homo sapiens*, genetic changes spurred on by radiation would have a greater effect owing to the smaller populations. This could lead to speciation in less time. But again, I just don't see that happening prior to the surviving groups getting back together and interbreeding. Most of the mutations would be detrimental to survival, and it still would take a long time to select out any beneficial ones. Not a good way to make *Homo nouveau*, from many points of view.

Types of catastrophes

- Collision with extraterrestrial object
- Nuclear holocaust
- Supervolcano
- Climate change
- Ocean acidification
- Atmospheric chemistry change

Another kind of catastrophe is one that is already under way: global warming. The average temperatures on earth are increasing at an alarming rate. That is a fact. The past decade was the warmest in recorded history, as was the last year (2016). Of course, recorded history only goes back a couple of hundred years. In the longer history of earth, there have been far greater warming periods than this one. At the current rate of warming, the average temperature on earth could rise four degrees Celsius by the end of the century. This will lead to melting of the ice caps and glaciers, sea levels rising up to seventy inches by 2100, acidification of the ocean, and changes in global ecologies. The results will be felt everywhere, but obviously most severely in regions near sea level. The resulting loss of food and habitat will most impact the poorest countries in the world.

As devastating as global warming will be on all the species of the earth, including *Homo sapiens*, no one is predicting our extinction because of it. It is not likely to lead to geographic splintering of our population, making allopatric speciation possible. In fact, it will probably cause more crowding of our current habitats. We will continue to occupy all land-based ecologies however crowded they may become, whatever warfare may ensue, and whatever starvation may occur. There will be no barriers to interbreeding, and there will be no speciation because of it.

The same is true of other types of slower catastrophes. We have had many other global changes in conditions that have had major impacts on evolution. Many warming and cooling periods have been well documented; these usually correspond to higher and lower carbon dioxide and methane levels of greenhouse gases. Changes in sea concentrations of hydrogen sulfide and other chemicals have also been well documented. All of these have dramatically changed the flora and fauna populations. Probably the most significant changes were in the amount of oxygen in the atmosphere and seas. We started with no significant oxygen in the early years of life, prior to the emergence of photosynthetic organisms and plants. In fact, oxygen was toxic to the early bacteria and archaea. Subsequently, we have had highly fluctuating levels. Today we think of our 21 percent oxygen level in our air as a constant, and all species living today are well adapted to this level. It has been far from constant, however. Fluctuations to as low as 10 percent and to over 30 percent have been documented in various previous periods, and each fluctuation was associated with a marked change in the fossil pattern.[183]

Any of these global changes can and probably will happen again over millions of years. Although they certainly can affect the number of *Homo sapiens* populating the planet and the available habitable locations, as will be discussed in later chapters, I think these are very unlikely to be the first forces leading to *Homo nouveau.*

In fact, I think it is unlikely that any catastrophic event will be the mechanism leading to *Homo nouveau.*

Chapter 11
Natural Selection

"The possibilities are limited only by imagination."

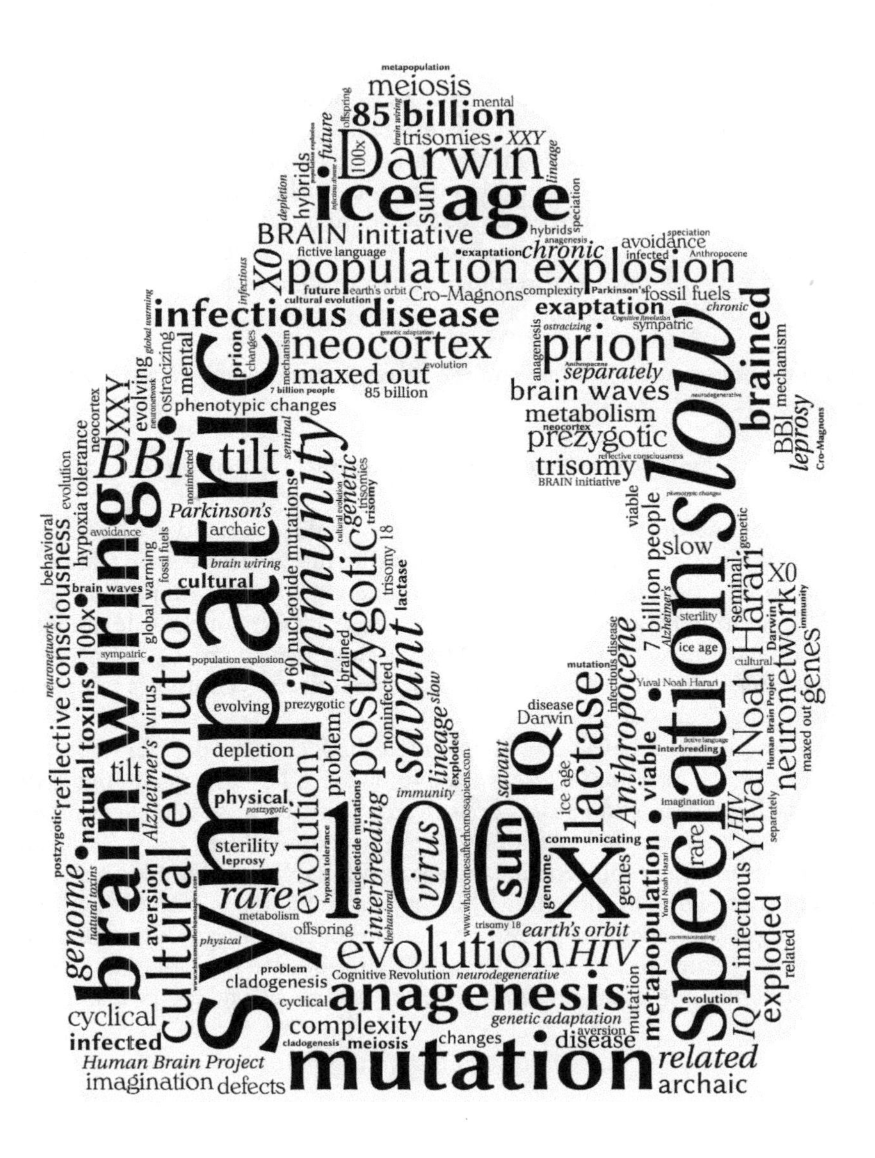

Now we're talking about speciation the old-fashioned way: by slow evolution and natural selection rather than a more dramatic catastrophe. There were about five hundred thousand years between the emergence of *Homo heidelbergensis* (or some other immediate ancestor) and *Homo sapiens.* Our speciation event happened in Africa, most likely by allopatric speciation. The particular differences between our immediate predecessor species and us were first determined by the random genetic changes during meiosis and by mutation. These created new phenotypes that were acted upon by natural selection. We have now gone another two hundred thousand years since *Homo sapiens* emerged and about fifty thousand years since we took over the world (at least in our own minds). What has happened in the interim with regard to this process?

One thing we know for certain is that our population exploded. We went from a few million individuals to over seven billion in just fifty thousand years! Something in our genetic makeup is working right—or wrong, if you're a Malthusian. Nothing like this has ever happened in our lineage before, so we are charting new territory with regard to our evolution. The other thing we know for certain is that we now occupy virtually every land-based, livable niche on earth. And we're all one interbreeding species, one big metapopulation; there is no room for allopatric speciation.

If *Homo nouveau* is going to emerge through natural selection, it will be by sympatric speciation.

Are we still evolving?

Our genes and the phenotypic changes they produce are, indeed, still evolving. In fact, there is evidence in our genome that humans have evolved one hundred times faster in the past forty thousand years than in all previous periods.[184] It is estimated that each child has as many as sixty nucleotide mutations passed on from its parents' DNA.[185] Meiosis continues to mix up the gene combinations. There is no question that natural selection continues to operate on these genetic changes.

For example, there have been documented changes in the past ten thousand years in the gene for lactase needed to digest milk. Historically this gene switches off once the child passes breastfeeding age, because in ancient times, we no longer ingested milk. This was before we domesticated cows. Today, 80 percent of Europeans still generate lactase as adults,[186] and this allows these adults to continue to drink milk without a problem. Similar recent beneficial gene changes have occurred in other digestion-relate genes as well as immunity genes, genes related to hypoxia tolerance at altitude, metabolism of natural toxins, and many others.[187] Natural selection and genetic adaptation are not optional.

Are we still one species?

In my mind, this raises an interesting point about speciation discussed in chapter 4 (see figure 9). Clearly *Homo sapiens* of today is a lot different from *Homo sapiens* of two hundred thousand years ago. That is, anagenesis has occurred within our lineage. But because we're not separately evolving from any existing descendant of our immediate common ancestor, we don't declare a new species. Had a group of earlier *Homo sapiens*—say, the Cro-Magnons—remained isolated from a group that has evolved into today's *Homo sapiens,* it is likely that we would have called us two different species by cladogenesis. That is not to say the Cro-Magnons would have stopped evolving; it's simply that they would likely have evolved differently. We don't name a new species unless it overlaps with some descendant of an immediate common ancestor, even though evolution and natural selection continue unabated. So, yes, we are still one species, *Homo sapiens*, even though we have evolved over the past two hundred thousand years.

How might we get to Homo nouveau by natural selection?

To review, in order to get to *Homo nouveau* by Darwinian evolution, two things must happen: selection of a separately evolving gene pool and the development of a barrier to interbreeding between the group of us in that gene pool and the remaining *Homo sapiens*.

The separately evolving gene pool could arise in two ways: new genetic combinations emerging through meiosis and mutation and/or a change in the environment that selects out some currently existing variation in our gene pool.

The possibilities for how all this could happen are too numerous to mention and entirely speculative. Just for illustration sake, I will mention a few possible paths that I have considered. The possibilities are limited only by imagination, and I will invite others to speculate on alternatives on the blog on my website, www.whatcomesafterhomosapiens.com. I'll start with a couple of low probabilities.

Chromosomal Error

What if somehow we changed the number of chromosomes we have? That number has changed many times over the eons, most recently when we evolved from a common ancestor with the apes with 24 pairs of chromosomes to humans with 23 pairs. We already have a number of existing trisomies (having three copies of a particular chromosome rather than the normal two) in today's humans, including Down's syndrome (trisomy 21), Kleinfelter's syndrome (XXY), Edward's syndrome (trisomy 18), and others. Turner's syndrome (X0) has one less chromosome. There are many other variations in the human chromosome number. If any of these were separately evolving metapopulations, we would already have a different species today. That is not the case. These abnormalities are rare and always have mental or physical defects leading to sterility, prezygotic avoidance, or otherwise less reproductive viability. Not my first choice for getting to *Homo nouveau.*

Environmental Changes

First, let me say that our biggest environmental threat today, global warming, is not high on my list as an environmental change that will lead to speciation. Although it will cause great damage and stress, I don't see a path to *Homo nouveau* as a result as discussed in the previous chapter. What about another ice age? Even though we are in the midst of a global warming epoch, scientists still predict that there will be another ice age in the future of our planet at some time, related to the eventual depletion of fossil fuels, and certain cyclical events related to the sun and the earth's orbit and tilt.[188] We survived the last ice age about fifteen thousand years ago, and I suspect the same would happen when the next one occurs. Although adaptive changes to cold climate as a natural selection force will occur, this is also low on my list of probable factors for speciation.

Higher Probabilities

Higher on my list of possibilities would be emergence of some kind of chronic infectious disease problem, particularly a virus or prion, for which there is no cure and which becomes epidemic worldwide. A prion is a short protein that can infect people and cause disease. A prion causes mad cow disease and also Creutzfeldt-Jakob disease, a neurological disease in humans. More recently, prions have been implicated as a possible cause or factor in Alzheimer's disease, Parkinson's disease, and other neurodegenerative illnesses.[189] People with an incurable chronic infection could still live for considerable periods of time and still have viable offspring who could be infected from birth or shortly thereafter.

How could a pandemic infection lead to *Homo nouveau*?

I can think of two possible speciation scenarios resulting from this. The first is that those not infected would avoid interbreeding with those infected (a prezygotic sympatric mechanism). Over time, then, the two metapopulations would diverge genetically to the point where some other barrier would reinforce the separate lineages. Does HIV qualify in that regard? I don't think so. With current treatment, the virus is not passed on to the offspring. Also, the avoidance of interbreeding between infected and noninfected individuals is not strong enough to allow separately evolving genetic pools.

What we would need is an infection that is always passed on to the offspring or easily contracted early in life, as well as one that produces a very strong aversion to interbreeding between noninfected and infected individuals. The only disease in history that may have come close to meeting at least the aversion criterion was leprosy. Centuries ago, lepers were kept in colonies for their entire lives and over generations. Their isolation was probably not long enough for a speciation event, and to my knowledge, we've never had a *Homo leprosarius*. Although we've never had an infectious disease that meets the criteria I've outlined, I would not rule it out entirely.

The second speciation scenario related to a pandemic infection would be if a portion of the population develops natural immunity to the infection. That, in itself, would not lead to speciation as long as those with immunity interbred freely with those without it. But what if the infected population or the immune population would develop some other genetic change such that the hybrids are not reproductively viable (a postzygotic sympatric mechanism)? Although this requires at least two unlikely events (the pandemic infection and the genetic change leading to sterility), I would give this scenario a greater probability. Speciation usually requires multiple unlikely events occurring over long periods of time. Also, this scenario does not involve behavioral selection or ostracizing of the infected population.

The Brain

With my imaginative *Homo sapiens* brain, I could describe many more possible scenarios that might lead to *Homo nouveau*. It is that same brain that in my view opens up the possibilities for evolving a new species. My bias is that any future speciation by natural selection would most likely be related to something changing in the way our brain is wired or functions. This assumes that evolution will proceed more or less the way it has for the foreseeable future. That's an assumption I will challenge in the next chapter, but first I will discuss some brain possibilities in more detail.

Looking at the indirect evidence, our brain has evolved since we've become *Homo sapiens*. By indirect evidence, I'm referring to the cultural evolution that has occurred in our species, viz., the behaviors we exhibit similar to those shown in figure 15 that distinguished the early humans. There is a mutual relationship between genetic evolution of the brain and cultural evolution, but it is not necessarily a direct causal relationship and not always obviously in synchronization. However, the cultural changes probably reflect at least some genetic changes in the brain.

With regard to our cultural evolution, there have been numerous so-called cultural revolutions in our history. Certainly the use of tools and fire, which preceded *Homo sapiens*, were seminal behavioral changes. Yuval Noah Harari, in his book *Sapiens*,[190] describes four cultural revolutions in the history of *Homo sapiens*.

1. The Cognitive Revolution: seventy thousand years ago, characterized by the emergence of fictive language. Many animals such as elephants and birds can communicate verbally, but we believe that only *Homo sapiens* can communicate about fictitious or imaginary concepts. This allows us to create metaphors that are very useful in communicating future plans, creating societal organizations, and in motivating others to act collectively in our interests. It allows us to talk about things that don't have a tangible objec-

tive reality. For example, discussion of religion is certainly enabled by fictive language, as are concepts like corporations, governments, countries, and even species. This greater cognitive ability was thought to be a key factor in our dominating all other human species.

2. The Agricultural Revolution: twelve thousand years ago. This is characterized by the domestication of plants for farming and animals for food, materials, and work. It changed us from peripatetic hunters and gatherers to occupants of stable villages, towns, and later cities. Collaborative groups became larger. Food became more abundant, although the variety of food in our diet decreased. Greater food abundance supported greater population growth and fed our high energy-consuming brain.
3. The Scientific Revolution: five hundred years ago. Humans began to study the heavens and explore the earth with a more systematic and scientific mind-set. Huge land discoveries were made and explored, and their biology and geology were studied and documented. Empires were built, trade routes were established, and a global economy began. Astronomy, archeology, geology, botany, and other sciences developed a more scientific method.
4. The Industrial Revolution: two hundred years ago. The steam engine and the internal combustion engine were invented. Textile manufacturing became automated. Trains, automobiles, tractors, and steamships proliferated. Electricity was harnessed. Organic chemistry and pharmaceuticals became widespread, including the conversion of nitrogen to fertilizer and explosives.

Most would probably add the Information Revolution and the advent of the Internet in the past two decades to this list.

Is our brain evolving?

Our behaviors as a species have certainly evolved rapidly. However, an archeologist a million years from now would not likely notice any change in the direct *Homo sapiens* fossil record during this same period. There would be no claims of new species that would correspond to these behavioral changes even though the evidence for the behavioral changes would be very apparent in the indirect fossil record. Just think about all the relic tools they will find: plows, cars, rockets, planes, computers, and cell phones, not to mention concrete, plastics, and radioactive fallout. Some are calling this a new geologic epoch—the Anthropocene.[191]

I believe that all of these cultural revolutions represent the same thing: advances in the way the brain functions. But unlike the brain changes in going from chimp-like to human, the brain changes of *Homo sapiens* have not involved enlargement of the brain. In fact, the size of the human brain has decreased slightly in the past twenty thousand years![192]

Instead, in my opinion the changes are in the wiring of the brain—the way the neurons connect. These wiring changes do not necessarily involve any genetic changes although genetic changes involving the brain are certainly occurring. That is, our rapid cultural evolution occurs at a pace far faster than genetic evolution. Cultural changes will be reflected only in the indirect fossil record years from now—not the direct fossil record. Because the brain does not fossilize, brain-wiring changes will not show up in the direct fossil record years from now either. Neither would other changes that could reflect natural selection show up in the fossil record, such as immunity to pathogens, metabolic and hormonal accommodations to urban crowding, processed food, stress, TV, the Internet, and other environmental forces. Nonetheless, as I said before, we already have become a different species from the *Homo sapiens* of two hundred thousand years ago by anagenesis. The question before us is how cladogenesis of *Homo nouveau* might happen.

Our Current Understanding of the Brain

The human brain remains one of the last frontiers in our understanding of human biology. Concepts such as the mind, thinking and consciousness remain poorly understood. The idea of a soul is more of a religious concept than a scientific one, but if it exists it would likely have something to do with the brain. Lots of unknowns—but lets start with what we know.

Our genes determine our brains' structure, wiring, and function. The structure of the brain is complicated. It reflects billions of years of evolution beginning with simple neuron clusters in worms to the highly complex and integrated structure of humans (see box). Our large neocortex is the *sine qua non* of *Homo sapiens.* It consists of a complicated, convoluted structure of nerves and supporting cells with deep grooves and high ridges. The detailed structure varies from individual to individual like a fingerprint. Three-dimensional imaging studies comparing this detailed structure in twins and nontwins demonstrated years ago the strong influence genes have on this structure.[193]

It is at the nerve networking level, as well as the microscopic and chemical levels of the brain, where the true understanding of the function of the brain will occur. Our knowledge at this level is woefully inadequate today. The concept of the three evolutionarily based components of the brain described in the preceding box (reptilian, mammalian and neocortex) is overly simplistic and misleading. "Higher level" functions are much more integrated throughout the brain.

Hopefully, our understanding will grow rapidly in the coming decades with the increased focus and funding of major research initiatives in both the US and Europe (see box). The goal is to fill in many of the gaps related to brain

Major components of the human brain

- Brainstem: also called "reptilian" brain or hindbrain—controls the most basic functions of life: cardiac functions and respiration. Also includes the cerebellum, a cauliflower-like structure related to muscle coordination and some higher-level functions as well.
- Limbic system: also called mammalian brain—controls hunger, thirst, fear, pleasure, other emotions, and long-term memory.
- Neocortex: largest in humans—highest levels of cognition, language, motor and sensory functions, and highest levels of thought and decision-making.

Brain research initiatives

- United States: BRAIN—Brain Research through Advancing Innovative Neurotechnologies. Announced by President Obama in 2013, this is an initiative involving the National Institutes of Health and four other federal agencies, along with numerous private research institutions.[194]
- Europe: Human Brain Project—funded by the European Union.[195]

function in a manner analogous to the federally funded Human Genome Project.

A very brief primer on how the brain works

(You can skip this section if you like—all you really need to know is that the brain is complicated.)

The basic cell in the brain that does its work is called a neuron. A neuron, or nerve cell, receives signals from other neurons and transmits signals to other neurons. Its input side, called dendrites, looks like a tree or bush with many branches connected to many other neurons to receive their stimuli. The output side, called the axon, connects to a smaller number of other neurons to transmit its signal.

The connections between the neurons are called synapses. Synapses are narrow, fluid-filled spaces between the neurons in which chemicals are released by the axon and are sensed by the dendrite to initiate an electrical impulse in the receiving neuron. These chemicals are called neurotransmitters. There are over one hundred different neurotransmitters in the human brain. Likewise, there are at least several hundred different types of neurons in the brain differentiated for special purposes.[196] In fact, in the human brain there may be thousands of different types of neurons, depending on how they are classified. They vary by shape, size, number of dendrites and axons, neurotransmitters, and the epigenetics or gene expression regulators active in the neuron.[197]

In addition to neurons, there are a variety of supporting cells in the brain. Recent evidence suggests that even these cells play a role in some cognitive functions. A typical human brain has about eighty-five billion neurons and about one hundred trillion synapses. Some of these synapses cause excitation of the next neuron, and some inhibit excitation of the next neuron. This is far more complexity than is contained in any computer. It is the way these neurons get connected and form various networks in the brain that determines how the brain functions. The pattern of these networks grows and changes continually from birth to death, the largest changes occurring in the first two decades. The neurons, synapses, neurotransmitters, and connections are all under genetic control.

The earliest studies of the brain basically depended on observing patients with specific neurological or mental problems and then examining their brains at autopsy to see which part of the brain appeared damaged under the microscope. This was later supplemented by brain stimulation of live patients with probes during neurosurgery. It is only in the last decade that our imaging technology, using MRIs and other technologies, has enabled us to examine the pathways in our brains associated with any mental function. These enhanced MRI technologies detect the flow of blood and water in the brain as specific tasks are being carried out. These flows reflect brain activity. This enables the creation of beautiful 3-D images of specific brain networks and pathways allowing us to actually "see" the brain working. The great thing about these imaging techniques is that they cause no harm (like x-ray radiation) to the body and are noninvasive. Their limitations today are the high cost and the somewhat slow time resolution. Our techniques are rapidly improving, and it is a matter of time before we know orders of magnitude more about how the brain works, particularly with the highly funded initiatives described in the box. Michio Kaku summarized our current technologies in his book *The Future of the Mind*.[198] Since the book was published in 2014, it has already become out of date! (See appendix 7 for a description of tools to study the brain.)

We have learned so far that the human brain is more complicated than any tools *Homo sapiens* have ever invented, including our most complex computers. Activities are going on in many places in the brain simultaneously, regardless of what is going on at the "conscious" level. I have *conscious* in quotes because we still don't even know what that means exactly. Many decisions, or at least acts, occur without our deciding it at the conscious level. There appears to be many different specialized centers of neural activity. There is evidence that many of the acts we perform that we think we are consciously deciding to do are reflected in neural activity before we consciously are aware of them.[199]

On the other hand, we are aware of some things, like emotions, where it is unclear why we experience them. Why do we experience fear, or anger, or love? One could argue that fear, for example, is a safety mechanism that causes us to run from danger. But as Yuval Harari discusses in his book *Homo Deus*, all of the mechanisms that lead to flight from danger happen at the subconscious level automatically: sensing the danger, stimulating the adrenal gland to pump adrenalin into the blood, increasing heart rate, activating muscles for flight, running to safety, etc. These subconscious reactions seem to be sufficient to protect us. Why experience the emotion of fear consciously? What was the selective advantage of experiencing the emotion? Maybe there is none and fear is just a brain phenomenon secondary to the same gene changes that causes the subconscious reactions. Perhaps it is just excess baggage. We don't know.[200]

In chapter 7, I concluded that there must be something different about the way our brain is wired that differentiates us from other species. In chapter 8, I explored what we know about our genome that seems to make a human different from a nonhuman. The answer seemed to be in our epigenome—the way genes are expressed. But these don't really answer the question. What is the impact of that wiring and genetic expression in what the brain can do? Is there a phenotype that characterizes the human brain?

What differentiates the human brain from all others?

One common answer is our ability to create and use tools. But so can crows and chimpanzees. Certainly our tools are more advanced, so is it that we do the same things but at a more complex level? Another common answer relates to language. Whales, birds, and elephants all have languages. Ants and bees have a sort of sign language. Is it simply that our language is more sophisticated in some way? Earlier in this chapter, we talked about fictive language. As far as we know, humans have been the only species (plural) that communicate about things that don't have an objective reality, like religious concepts, imaginary animals, and organizations. Multiple *Homo* species have had that ability as evidenced by wall art, carvings, and other artifacts. Do any nonhuman species imagine similar things? We have no way of knowing, but a crow must somehow imagine a tool that it is about to create that doesn't yet exist. Our fictive language implies fictive thinking as well. Does a crow have fictive thinking without the language? It is difficult to define exactly what our brains are doing that other brains don't do—that is, it is difficult to define our unique brain phenotype.

John Hands, in his book *Cosmosapiens*, spends considerable effort to try to nail this down. It is certainly more than consciousness defined as an awareness of both self and surroundings; many species are conscious by that definition. His conclusion is that *Homo sapiens* have a special kind of consciousness, which he calls reflective consciousness. He defines reflective consciousness as "the property of an organism by which it is conscious of its own consciousness, that is, not only does it know but also it knows that it knows."[201] We are the only species that tries to understand what we are and where we came from, and that asks such questions as "What comes after *Homo sapiens*?" We not only experience emotions, but we know we are angry, or sad, or fearful. It is somewhat unclear from his research whether or not reflective consciousness existed in other *Homo* species like Neanderthals, and when exactly this brain phenotype emerged, but there is no question in his mind that today *Homo sapiens* is the only species with this capability. What we don't know is exactly what in our genome is responsible for this capability, or how exactly the brain wiring creates it.

Michael Tomasello and Malinda Carpenter explain another possibly unique human phenotype called "shared intentionality." In studies of chimpanzees and infant humans (prior to language development) they showed that both chimps and humans have the ability follow the gaze of an adult human to attempt to see what the adult is looking at. Only the human infant, however, uses gestures of various types to try to share that information with other infants. In this and a series of other experiments with both chimps and infant humans, they demonstrated an attempt to elicit collaboration of various types only in the humans. They concluded

> *The emergence of these skills and motives for shared intentionality during human evolution did not create totally new cognitive skills. Rather, what it did was to take existing skills of, for example, gaze following,*

manipulative communication, group action, and social learning, and transform them into their collectively based counterparts of joint attention, cooperative communication, collaborative action and instructive learning—cornerstones of cultural living. Shared intentionality is a small psychological difference that made a huge difference in human evolution in the way that humans conduct their lives.[202]

According to Steven Sloman and Philip Fernbach in their book *The Knowledge Illusion: Why We Never Think Alone,* this shared intentionality was key to early humans being able to hunt larger prey like mammoths and bisons so effectively. Clearly other animals hunt collectively, but the early humans' ability to overcome their physical limitations so successfully was related to their complex social skills in planning an executing a collaborative hunt. Sloman and Fernbach use similar phraseology in describing shared intentionality as John Hands used in describing reflective consciousness: "...we know that they know that we know that they know, etc."[203]

If shared intentionality is indeed a key differentiator of the human brain phenotype, all that we can say from these experiments between today's chimps and humans that that this phenotype developed sometime in the period after we split from the chimps 5.4 million years ago. Whether the earliest pre-humans such as Ardi or Lucy, or even the earliest humans such as *Homo ergaster* or *Homo erectus* had this brain phenotype is unknown. However, there is strong evidence that social and cultural complexity occurred in direct proportion to the size of our neocortex, as reflected in the increasing size of our groups and communities.[204] Thus it seems likely this brain phenotype developed *after* we became humans as our brains grew in size.

As complex and even mysterious as the current human brain seems, I am confident we will figure it all out. We do not have to evoke supernatural forces simply because it is currently beyond our understanding. We are at a similar place with the brain to where we once were with DNA, genes, and the genome. The late Carl Sagan put it well when he said in his book *Dragons of Eden*, "My fundamental premise about the brain is that its workings—what we sometimes call the 'mind'—are a consequence of anatomy and physiology, and nothing more."[205] Unlike species, there is an existent reality to the human brain waiting to be discovered and understood. In spite of our ignorance, I can still speculate on how a brain change can lead to *Homo nouveau.*

Possible speciation related to brain evolution

You will note that I'm focusing on the brain and not heart, liver, muscles, immune system, or other parts of our bodies that will continue to evolve as well. Our brain got us to *Homo sapiens* and, in my view, it will likely be what gets us to *Homo nouveau* assuming it happens by natural selection.

It will be some phenotype change in the brain on which natural selection will act. That is, there will need to be something about the way the brain works that conveys some competitive reproductive advantage to the new species. That could be better functioning of the brain in the way that fictive language worked millennia ago, or it could be resistance to neurological disease like Alzheimer's, or some other function.

Multiple human brain phenotypes already exist.

What we do know is that the human brain is evolving. At least one-third of all of our genes are targeted primarily to the brain![206] That is more than is devoted to any other organ. These genes are being reordered by meiosis and are mutating just like any other genes. Most of these changes don't do anything. Some of those changes are causing harm, like Alzheimer's disease. Because Alzheimer's disease occurs after peak reproductive years, negative natural selection is much weaker. Some mutations are being selected because they do something good. They are reflected in the cultural revolutions already discussed. They are reflected in the HARs discussed in chapter 8. That's how we got to this point in the first place.

Are there variations in human brains today? Absolutely! I don't want to get into a debate about intelligence regarding what it means and how it is measured. You can decide for yourself whether you think some people are smarter than others. There is no question in my mind, however, that some people's brains function differently from mine. Being able to do higher mathematics or having musical ability are two examples that I believe require some special brain wiring that I don't have.

Evidence of Brain Wiring Variation

There was a series of experiments done by Walter Mischel and his associates that wonderfully illustrate the relationship between phenotypic brain behavior and brain wiring. They took children at age four and put a marshmallow in front of each one. The children were told they could eat the marshmallow now, or if they waited for some period of time, they could eat two marshmallows. The researchers then left the room and observed whether or not the child ate the marshmallow or waited. Some ate immediately, and some waited to get two marshmallows. They then did follow-up studies on these children up to forty years later! The group of children who at age four managed to delay gratification in order to get a bigger reward later consistently "developed into more cognitively and socially competent adolescents, achieving higher scholastic performance and coping better with frustration and stress."[207]

A subset of these same individuals underwent functional MRI studies during various visual stimuli forty years later, and they demonstrated distinct differences in brain circuit reactions to those stimuli between the delayed gratification group and the non-delayed-gratification group.[208] There was something about the way these two groups of brains were wired that allowed one set of four-year-olds to delay gratification and not the other group. This wiring change was demonstrable forty years later when functional MRI became available. Whether or not you call this brain difference intelligence or something else, clearly the way the brain is wired affects social behavior and success in life later on. Not all human brains are wired the same way.

What about the "savants" whom we know exist: people who can remember everything that happens to them, or people who can multiply any two five-digit numbers in their head quickly, or people who can tell you instantly the day of the week for any date in the past or future? On March 15, 2015, Rajveer Meena, an Indian student, correctly recited pi to seventy thousand decimal places over a ten-hour period.[209] Surely something is going on in his brain differently than most other people.

There is a syndrome, called Savant Syndrome, which was popularized by Dustin Hoffman in the movie *Rain Man*. Savant Syndrome is characterized by some mental disability (usually autism) combined with some extraordinary mental ability in art, music, math, or memory. The person that inspired *Rain Man,* Kim Peek, had an extraordinary defect demonstrated in images of his brain in that the main connections between the right and left parts of his brain were missing. Obviously his wiring was really different. Another famous person with Savant Syndrome, Daniel Tammet, describes in his book *Born on a Blue Day*, how he thinks. Words and numbers appear to him as "shapes, colors, textures, and motions."[210] To him, Wednesdays have the color blue. There is something about the way the neurons in the brains of these remarkable people are connected that allows them to do those seemingly impossible mental feats that is different from the way the neurons in most brains are connected.

There is another syndrome, Highly Superior Autobiographical Memory (HSAM), in which certain adults can remember virtually every detail of their own lives, knowing the exact date and day of week in which events related to them happened and all the details surrounding those events. Their memories, however, are not unusual regarding things unrelated to their personal lives. Unlike Savant Syndrome, they don't have autism or other mental disorders. MRI studies show significant differences in the connections in their brains compared to "normal" controls.[211]

It is clear that *Homo sapiens* of today have tremendous variation in our brains. We can measure the difference both in the genes that we know are related to the brain, as well as the brain structure as shown by imaging techniques. Does this mean that *Homo nouveau* already exists? Do people with extraordinary brain functions already constitute a different species from *Homo sapiens*? I don't think so. They are not separately evolving metapopulations. All of the people with brain variants today interbreed with all *Homo sapiens* freely and produce viable offspring that also interbreed freely.

Now for some fictive speculation.

Looking ahead to the next million years, what are some of the things we might expect in brain evolution and natural selection that could lead to cladogenesis of our lineage?

Brain waves.

For a neuron to fire, a chemical reaction needs to occur across the membrane of the neuron cell at the synapse. A "hole" opens up in the membrane, and positively charged ions rush in, causing a spike in the electric potential of the membrane at this point. This electrochemical spike is then propagated from the dendrite to the neuron body and finally down the axon, where it terminates at the next synapse. That's how nerves work.

When many neurons in the same vicinity discharge simultaneously, it creates a larger voltage change in that part of the brain, which can be detected by sensors on the scalp. This is how an electroencephalogram (EEG) works. Basically, an EEG is picking up voltage waves in the brain related to brain activity. There are different types or patterns of waves generated by different types of brain activity in various locations. Interpreting these waves has been used clinically for decades in a variety of situations, including diagnosing epilepsy, sleep disorders, certain psychoses, and other neurological problems.

Recently, some amazing things are happening to these waves. A number of experiments have been reported demonstrating what is called BBI, or brain to brain interface. In one such experiment,[212] an EEG signal was used to detect whether or not a participant was responding yes or no to a question on a computer screen. The participant responded by either looking at one light signal with a specific light frequency representing yes, or a different light signal with a different light frequency representing no. The EEG signal was different depending on which light the participant was looking at. A computer was programmed to detect this difference and then transmit the answer over the Internet to an electrical apparatus adjacent to another participant located one mile away in a different building. It was this second participant who had generated the question to the first participant. The electrical apparatus next to the second participant generated what is called a TMS (Transcranial Magnetic Stimulation) signal[213] to the scalp of the second participant if the answer was yes, and not any signal if the answer was no. The TMS signal causes the second participant's visual cortex to be stimulated in a manner that allows him to know whether the answer was yes. In this manner, the two participants were able to solve problems together without any further communication. This demonstrated that with computer assistance, we can now silently communicate between two human brains by detecting brain waves in one person and providing TMS stimulation to the other. I'd call this an early form and artificial example of brain wave communication.

In an even more amazing experiment, using sophisticated computer programs, researchers can now tell from brain waves what a person is thinking—up to a point.[214] This is done by placing an array of electrodes directly on the surface of the brain to capture the brain wave signals. This requires invasive surgery, where a portion of the patient's skull is temporarily removed in order to surgically attach the electrodes.* Eliminating the need for the waves to traverse the bony skull allows the brain wave signals to be detected much more

*These experiments are performed on patients who must have their brains exposed for other medical purposes.

strongly. The signals are sent to a computer. After a period of "training" the computer program, the computer can then display the words that a patient is thinking—analogous to voice recognition.

This is mental telepathy without the fraud! The technology is called ECoG (electrocortography). Equally amazing, this same technology is beginning to be used to allow a patient who is paralyzed to control prosthetic devices entirely by thinking about the movements.[215-217] True telekinesis! A commercial company, Emotiv, now sells a device that allows you to control a computer, wheelchair, or other devices using a simple, noninvasive headset that detects your brain waves for its controls.[218]

Can brain wave communication lead to Homo nouveau?

What does this have to do with *Homo nouveau?* First, it demonstrates the degree of detailed information that is buried in these brain wave signals once our computer software technology has advanced enough to decipher the information. Second, we know the human brain is the most advanced pattern recognition tool that exists—far better than any computer program. There is no reason that the human brain couldn't do a better job of understanding human brain waves than a computer if there was some way to get the signals from other humans into our brain other than artificially using TMS or other tools. We already have the "software" for pattern recognition in our brains.

All it would take is a biological detector rather than the electronic one that ECoG or EEG uses. I know this is probably the most speculation that I have yet introduced into this discussion, but over the next million or so years, I believe this is entirely possible. We have evolved biological detectors for equally improbable signals like electromagnetic radiation used for sight, detection of air compression signals used for sound, and chemical detectors used for smell. Other species have developed magnetic field detectors for navigation. These biological detectors are typically superior to any artificial detectors we can build. For example, using dogs to search for missing humans in the rubble of an earthquake or explosives in airport luggage is the best smell detector available. A biological detector for brain waves is no more outlandish than any of those. To me it is just a matter of time and chance. We have both.

Sooner or later, the genes controlling some part of our body—perhaps some currently specialized nerve endings like the cones in our retina—may mutate in a way that reacts to brain waves. That's how things have happened with our other signal receptors including the ear and the eye discussed in chapter 4. It is exaptation.

I have no idea what mutation or mutations in what genes would allow exaptation of some existing nerve component to make the brain wave detector happen, or where the detector would be located on our body. It is pure random chance. Perhaps it will be some component of the retina, or the fingertips. Stranger things have happened to us than the emergence of a brain wave detector. I can assure you that once we get our computers out of the loop, BBI will be much more efficient and capable. Should such a mutation occur, it will likely be selected for future generations.

Keep in mind that the evolution to brain wave communication does not need to occur in a single step. Our current retina is the result of many different mutations over a long period of time. Initially, light receptors were crude, only detecting the presence or absence of light. Later, they got more refined to detect different wavelengths. Similarly, early brain wave detectors may be crude and have minimal resolution but could improve with time and subsequent mutations. Likewise, the strength of the brain wave signal might become enhanced through other mutations. Evolution is a process, not an event.

Imagine the benefits such a biological brain wave detector could convey to those with it. Communication could occur silently and without detection between individuals with this capability while in the presence of humans lacking it. Those with it could eavesdrop on the intentions of people with whom they are negotiating, or competing with, or even warring with. They could surreptitiously control devices. This could even happen over long distances by telekinesis of the signals to long-range communication devices. The implications are enormous and could convey a competitive advantage far beyond those of language and other cognitive capabilities that allowed *Homo sapiens* to outlive all the other *Homo* species.

It is already happening...almost.

If you think I'm now in the realm of science fiction, you need to have your brain rewired. The proof of concept is already occurring in the laboratory of a neuroscientist at Duke University, Dr. Miguel Nicolelis.[219] He has connected the brains of rats and also of monkeys into what he calls Brainets. He does this by inserting electrodes in the brains of the animals, connecting those electrodes to a computer, and then transmitting the signals back into the other animals' brains in the Brainet. He then allows them to cooperate in performing various tasks that result in rewards to the animals. In all cases, the collaborating animals in the Brainet perform the tasks either at the same or higher level than the animals not connected![220]

Although Dr. Nicolelis' Brainets do not yet use biological brain wave receptors, they are functioning in the same way I envision. In fact, I think the biological version, if and when it occurs, will be far superior to the artificial one that Dr. Nicolelis is able to build today. It will not be limited to only the nerve cells adjacent to a small number of electrodes used in the Brainets, and it will not have to go through the intermediary of our inferior computers.

If and when humans develop biological Brainets, they will be a superior group. The only question is whether this superior group would remain reproductively isolated from humans without these biological detectors. Would they choose to breed only within their kind as a form of sympatric behavioral isolation? The same question could be asked of any new brain variant I can dream up. We'll come back to this question shortly.

Have We Reached the Limits of What our Current Brain Can Do?

What other brain changes could lead to *Homo nouveau*? The possibilities seem limitless. Who would have ever thought we'd have people who could do the things the savants or HSAM people can do? Perhaps we'll see savants without the autism. I doubt one hundred years ago someone writing a book like this would have thought about reciting pi to thousands of decimal places. What about great musicians, artists, and mathematicians? Although it is fun to think up new ways of thinking, what already exists blows my mind. There is something about the model of the brain that we already have that is flexible enough to do anything conceivable regarding cognition—and even things that are inconceivable. It took the fourth most powerful computer in the world forty minutes to simulate one second's worth of human brain activity representing one percent of the human brain.[221] We aren't even close to building anything that can compete with our brain. To have parallel processing using billions of neurons connected in near infinite ways is an amazing recipe. "The human brain ... is the most complex piece of organized matter in the known universe."[222] Nothing would surprise me about the future. As Shakespeare said in Hamlet (and memorialized in *Hair*),

> *What a piece of work is man! how noble in reason!*
> *How infinite in faculty! in form and moving how*
> *express and admirable! in action how like an angel!*
> *in apprehension how like a god! the beauty of the*
> *world! the paragon of animals!*

There is no doubt that the human brain will continue to evolve in ways that will make today's *Homo sapiens* seem primitive. Natural selection will act on random genetic changes like it always has. Michio Kaku argues that one of the ways the brain will *not* change is to get bigger.[223] So far he has been right, at least for the last twenty thousand years. He thinks we are at the physical limits as to how far brain neurons can go to connect before thinking becomes too slow and energy consuming, or how thick neurons can become to increase speed (thicker neurons transmit faster), or how thin they can become to pack more in the same space without damaging their electrical properties, or how many more connections they can make without becoming too energy consuming.

I think he is largely raising straw men here. There is no evidence that we need any more neurons (with eighty-five billion of them) or more connections (with over 100 trillion of them), or that we're somehow maxed out in using what we already have. In fact, I think we already have evidence that if we change the way we use the neurons we already have, particularly the way they connect, we can still do a lot more amazing things with the physical brain we have.

Think about it. Modern *Homo sapiens* came on the world scene about 200,000 years ago. Our brains were about the same size (or even larger) than they are now. But what we can do with those brains 200,000 years ago pales in comparison to what we can do today. Then, we were only slightly more advanced than today's chimpanzees. We couldn't talk; we didn't wear clothing; we were still peripatetic hunters and gatherers living in small groups; and our tools were primitive. All of our cultural revolutions have occurred since then: cognitive, agricultural, scientific, industrial, and informational. We've built great cities, gone to the moon, have incredible tools, and most importantly, we *understand* how and why we do all these things. We've developed this highly advanced culture. All of this has been accomplished with roughly the same brain, the same number of neurons and the same "limited" neuron physiology. There is no evidence we have maxed out on what this amazing organ can accomplish.

Where's the barrier?

There will surely be superior brain changes ahead—probably mostly in the wiring. The only question in my mind is whether speciation of such a group with these changes will occur. We are at an unprecedented time in our genus lineage—the first time that we have occupied every livable, land-based ecological niche on earth. It is not just our occupation of every livable location, but also our ability to get anywhere on earth in less than a day. It would be pretty difficult to isolate a group of us for a million years, or even a few thousand years, to allow allopatric speciation to occur no matter what changes occur in our brain or in anything else. The one exception to that might be if we colonized an extraterrestrial body. That leaves us with sympatric speciation as the probable mechanism.

How might sympatric speciation happen?

Looking at appendix 3, there are many mechanisms of sympatric speciation, and none of them has ever applied to our *Homo* genus lineage as the primary mechanism of speciation. Certainly behavioral isolation may have been a factor preventing a lot of interbreeding between the Neanderthals and us, but that was not a speciation event. *Homo sapiens* originated by allopatric speciation in sparsely populated Africa far from the Neanderthals.

Nonetheless, behavioral isolation remains one candidate for creating *Homo nouveau* in the absence of geographic isolation. What if the people with some superior brain trait avoided interbreeding by choice with those without the trait? Or worse yet, what if they decided to "ethnic cleanse" everyone else? Although theoretically possible, we've had some form of voluntary isolation of one group or another, and even ethnic cleansing for millennia at least. It has not yet proceeded to the point where a barrier to interbreeding has occurred and, with the current population size and mobility, it seems an unlikely path to *Homo nouveau*.

There is a postzygotic mechanism that I believe is more likely: that eventually those *Homo sapiens* with the superior brain trait produce sterile hybrids

*A horse has thirty-two pairs of chromosomes and a donkey has thirty-one. A mule has sixty-three total chromosomes which means one of them is unpaired.

when mating with those without the trait. In that case, mating between those with and without the trait would occur normally. But they would not produce viable offspring. There would be no ostracizing, no cultural or racial wars, and no major social upheavals, at least in the immediate thousand years or so. But there would be no mixing of the gene pools either. Thus we would have two separately evolving metapopulations that would continue to diverge with new SCs, as shown in figure 19. We'd achieve cladogenesis into *Homo nouveau*.

Why do I think this is more likely? There is nothing about a mutation giving humans some superior brain trait that would also lead necessarily to sterile hybrids when mating with those without the trait. I'm suggesting that these are two separate, random mutation events that just happen to occur. One mutation (or set of mutations) leads to the superior brain trait, and the other mutation would lead to sterility of the hybrid offspring. If both of these mutations occur, there would not be strong negative natural selection to eliminate the combination. Because *Homo nouveau* would have such a strong positive brain trait, it would continue to be selected to survive. Remember, the offspring of two *Homo nouveau* would be viable. The situation would be the same as today's horses and donkeys, which continue to survive even though the mules they produce by interbreeding are sterile.* Figure 19 shows the de Queiroz diagram for this scenario.

FIGURE 19

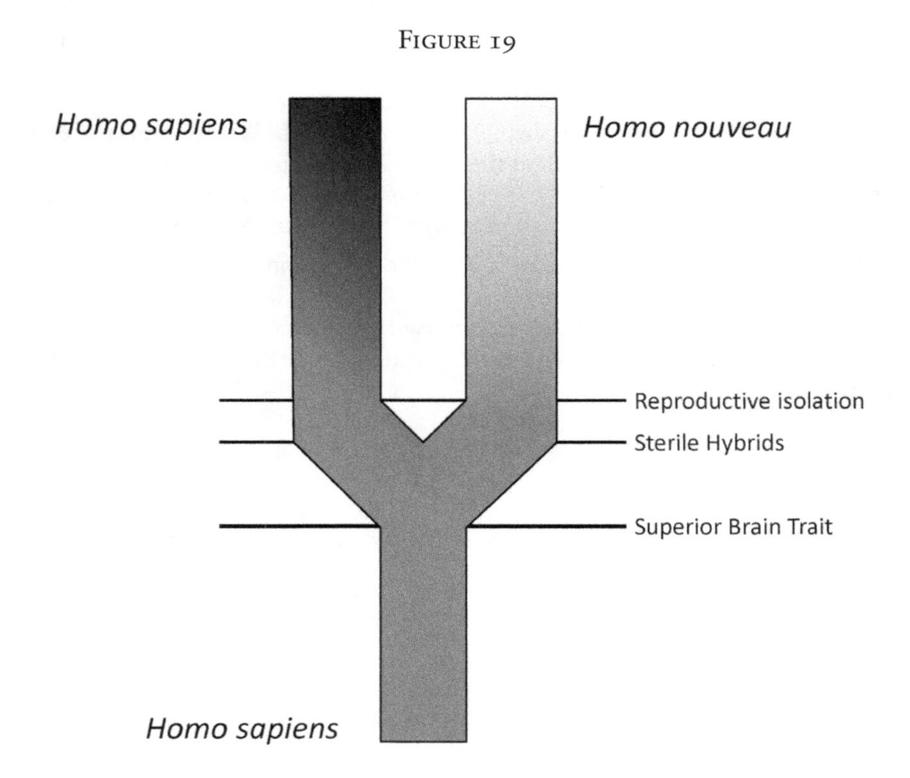

What's wrong with this picture?

Looking at figure 19, the main issue (and perhaps the main flaw) in this picture is the definition of "superior" regarding the brain trait that will lead to speciation. Remember, natural selection is aimed solely at increasing the ability to procreate, nothing else. Some of the brain changes I have been speculating about may have nothing to do with increasing procreation capability or survival in any way. What if we had a group of us that could do the things the savants do without the mental disorders that currently accompany them? Does doing higher mathematics or physics allow someone to succeed in ways that lead to greater numbers of them? As Michio Kaku points out in his book *The Future of the Mind*,[224] most people don't particularly want to do higher mathematics or theoretical physics, or aspire to professions where that is required. They also don't make more money by doing that. The most financially successful people in our societies seem to be athletes, movie stars, or businesspeople, none of which necessarily need the kind of brain rewiring I've been speculating about.

An inconvenient truth

That brings me to one final point that bears heavily on what human trait might be selected for speciation: wealth, or lack thereof. Much of what today's humans struggle over is how to succeed financially—how to raise themselves and their children out of poverty if they are poor, or how to remain in the middle class if they are already there, or how to get rich. We don't measure or particularly value success based on the number of children people have. If evolution and natural selection are the major forces shaping *Homo sapiens* today, we must reconcile this against two facts that characterize our species today.

1. We have an increasing disparity of wealth distribution. Fewer people are rich, and more people are poor.
2. Figure 20 shows a plot of countries on two axes: average wealth as measured by per capita GDP versus fertility rate. Clearly the poorer the country, the higher the birth rate.

FIGURE 20

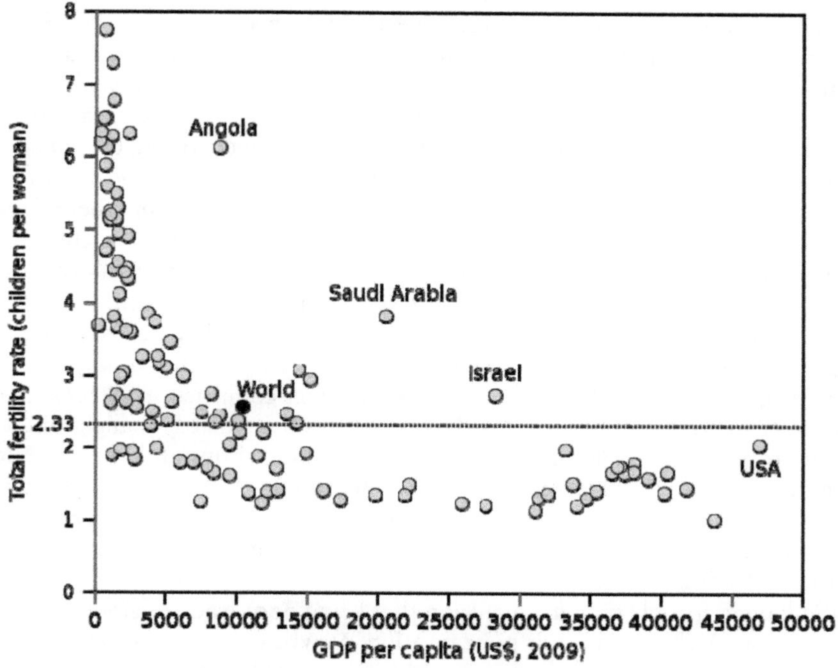

Source: Wikipedia: https://en.wikipedia.org/wiki/Demographic-economic_paradox

These facts seem to raise an interesting question: Could natural selection for *Homo sapiens* at this point in time favor being poor? This is not a new question. Ronald A. Fisher, the famous statistician involved in the modern evolutionary synthesis discussed in chapter 5, raised this same question close to one hundred years ago. He stated[225]

> *We must face the paradox that the biologically successful members of our society are to be found principally among its social failures, and equally that classes of persons who are prosperous and socially successful are, on the whole, the biological failures, the unfit of the struggle for existence, doomed more or less speedily, according to their social distinction, to be eradicated from the human stock.*
>
> *Numerous of investigations, in which the matter is approached from different points of view, have shown, in all civilized countries for which data are available, that the birth-rate is much higher in the poorer than in the more prosperous classes, and that this difference has been increasing in recent generations.*
>
> *The type of man selected, as the ancestor of future generations, is he whose probability is least of winning admiration, or rewards, for useful services to the society to which he belongs.*

Are there other explanations for Fisher's observations and for figure 20? Could it be that people in the wealthier countries have figured out that we have too many people on earth to survive and have been smarter at birth control? If anything, it is the governments of poorer countries like Rwanda that are actively pursuing child limitations. Perhaps the people in wealthier countries limit children to have more time to pursue activities that make money. That would suggest that our smarter brains are actively trying to reduce reproductive success, which heretofore has been the measure of success for natural selection. That is, have we come to an evolutionary point where there is a conflict between the historical result of natural selection and the cultural goal emanating from our highly cognitive brain, which was the result of natural selection in the first place? This is too speculative to say for sure.

Natural selection is not a conscious decision. Perhaps what we are observing is natural selection working the right way (reducing fertility to improve survival), and we're seeing its effect in the wealthier countries. The fact is that for 99.999 percent of the time that *Homo sapiens* has existed, changes in the brain probably dominated our ability to compete and thrive and increase our numbers. It took 200,000 years for the population of *Homo sapiens* to reach one billion, but it then took only another one hundred years to reach six billion. This exponential explosion in growth rate is dramatically changing our natural selection forces.

I'm having trouble with a conclusion that being poor leads to greater reproductive success. Correlation is not causation. Maybe causation goes the other way around: great reproductive success in our case led to being poor. My brain is telling me that somehow being smarter or having superior brain functions is still the key to survival of *Homo sapiens* in the future. It certainly was in the past.

Is intelligence still being selected?

However, the evidence, at least in one study in the United States, is that higher intelligence as measured by IQ tests correlates with lower birth rates. This effect seems to be mediated through higher education, which also correlates with lower birth rates.[226] The prediction from this study is that "IQ will decline by about 2.9 points/century as a result of genetic selection," and further, "the proportion of highly gifted people with an IQ higher than 130 will decline by 11.5% in one generation and by 37.7% in one century." This study was consistent with many other studies cited elsewhere.

This does not bode well for the future of *Homo sapiens*. Currently, natural selection seems to be favoring less intelligent humans having higher birth rates than more intelligent humans. The projections are that human population growth is continuing apace, projected to reach eleven billion by 2100 mostly owing to increases in poor countries.[227] This is the first time in our 200,000-year history that the human population has gotten to a tipping point. We are running out of natural resources to sustain our exponential growth. We are running out of drinkable water, food, land, and maybe even oxygen with the amount of deforestation going on. The sustainability think tank Global Footprint Network estimates that humans today consume resources at a rate that would be required of 1.6 earths.[228] Obviously that's not sustainable. In the past, if natural resources were scarce, there was always someplace else to try. We no longer have that option, at least on earth.

The vast majority of species that ever existed are now extinct. Natural selection does not lead to the survival of most species—only the most fit to survive given the immediate environment. We may not be one of those species going forward. Somehow envisioning a future world dominated by impoverished, less intelligent humans isn't where I thought all of this was heading. But that is teleological thinking rather than natural selection thinking.

Evolution and natural selection do not work from a script or plan; they largely depend on random changes to the genome. Although natural selection does tend to maximize certain traits, that maximization process could lead to dead-ends or a condition where further maximization is ineffective or even harmful. Take, for example, body size or strength. Probably larger and stronger hominids were beneficial traits in going from the relatively small Australopithecines to the larger and stronger Neanderthals. However, once humans learned how to kill at a distance using spears, then bows and arrows, and finally bullets and missiles, size and strength no longer are survival advantages. That is one reason why the smaller *Homo sapiens* are still here and the Neanderthals are extinct. The same phenomenon could be happening with intelligence.

Perhaps we have maxed out on its survival advantage.

Chapter 12
Genetic Engineering

"We may not want to wait for natural selection to create *Homo nouveau.*"

Given the ending of the last chapter, we may not want to wait for natural selection to create *Homo nouveau.* "Man is not going to wait passively for millions of years before evolution offers him a better brain" said Romanian psychopharmacologist Corneliu Giurgea. He studied drugs to enhance human cognition.[229] Why use drugs when we now have a more powerful tool: genetic engineering?

We have used genetic engineering for millennia—only we didn't call it that.

Homo sapiens has been genetically modifying plants and animals for at least twelve thousand years. We call it selective breeding. That's how we got all those breeds of dogs and big-breasted turkeys that can't fly and succulent Golden Delicious apples. Even before they knew how everything worked at the gene level, our ancestors knew that if they selected the biggest or the sweetest or the whatever-est of a plant or animal and mated it with a plant or animal with a similar trait, the offspring would in most cases have that trait as well. Eventually, repeating this selective breeding over generations would lead to new strains of plants and animals and ultimately to new species. As discussed in chapter 4 on speciation, this is sometimes referred to as artificial selection. It is a slow and imprecise tool.*

In chapter 5, we outlined the history of our understanding of what was really going on with selective breeding (and in fact all breeding). Remember Mendel's peas? By the end of the twentieth century, we understood the chemical and microscopic makeup of genes, DNA, RNA, and chromosomes and how they worked. In fact, we learned how to bypass the slow process of selective breeding.

What is genetic engineering?

The amazing thing is that every plant and animal is *identical* in every way except one! Every cell in every plant and animal has a nucleus. Every nucleus has a full set of chromosomes consisting of DNA. The DNA has the same double-helix structure and consists of the same four nucleotides. The genetic code that translates the nucleotides into amino acids (remember the codon?) is identical in all plants and animals. Our proteins consist of the same twenty amino acids. *The only difference between us and a pea or a worm is the sequence of those nucleotides!* That accounts for all of our variations. Genetic engineering is simply the ability to alter that sequence directly in a laboratory.

Every cell in a plant or animal has a full set of genes (with one exception mentioned later). The genes are the targets of genetic engineering. The goal is to alter a gene or genes in at least some of the cells of an organism. That alteration could be in the form of replacing a part of a gene with a correspond-

*There is another process which has been used for decades called "directed evolution" in which novel enzymes are produced in the laboratory by inducing mutations and novel recombinations of genes of natural bacterial enzymes. These new enzymes are used for industrial and other purposes.

ing part containing a different nucleotide sequence, usually with the intent to remove a deleterious mutation. For the same reason, it could also involve removing or disabling an entire gene and not replacing it. Finally, it could be to add a gene or gene segment that does not exist in the cell to provide some function that's not already there.

How did we learn how to do genetic engineering?

In the early 1970s, Paul Berg and his colleagues at Stanford showed that it was possible to selectively take a section of nucleotides from the DNA of one organism and insert it into the DNA of another organism.[230] This was the first demonstration of the creation of what is called recombinant DNA, and he received the Nobel Prize for his work.

In 1973, Herb Boyer and Stanley Cohen experimented with common bacteria found in all humans, *E. coli*. These bacteria contain plasmids, which are small circular pieces of DNA separate from the chromosomes. Plasmids contain genes that produce proteins. They play a role in a small number of functions, including antibiotic resistance. Boyer and Cohen transferred plasmids from one strain of *E. coli* having resistance to a single antibiotic into another strain that did not have that resistance. The receiving strain subsequently had resistance.[231] This showed that the genes from one strain of *E. coli* could be transferred to another strain and function normally.

Even more significant was when they showed they could selectively cut out part of the plasmid with the resistance gene and patch it into the plasmid of another strain of *E. coli*, thus creating recombinant DNA with the resistance gene.[232] This is the key to genetic engineering—the ability to isolate a small segment of an organism's DNA that contains a particular nucleotide sequence of interest, make many copies of that DNA segment and then splice or patch that segment into some delivery vehicle (called a vector) to create recombinant DNA. If the recombinant DNA gene is from the same species as the target organism, the process is called *cisgenic*.

Boyer and Cohen didn't stop there. Next, they created a recombinant DNA plasmid that contained the genes for resistance to multiple antibiotics. They did this by patching genes from a different species of bacteria (*Staphylococcus aureus*) into a single *E. coli* plasmid and transferring that plasmid to other *E. coli*. The receiving *E. coli* not only became resistant to all the antibiotics but also conveyed those same traits on to future generations of that strain of *E. coli*.[233] This showed that genes from one species of bacteria could be transferred to a different species of bacteria and be incorporated into the genome of that target bacteria for future generations. In contrast to cisgenic (when genes are transferred between the same species), the transfer of genes between different species is called *transgenic*. Genetic engineering was now taking off.

To really make the transgenic point, they demonstrated that they could create an *E. coli* recombinant DNA plasmid that contained a gene from a toad! The toad gene functioned in all future generations of that strain of *E. coli*. This showed that transgenic engineering could occur virtually unconstrained by species and set the stage for human genetic engineering.

If you could transfer a gene through recombinant DNA from any species to any other species, the implications were enormous. First, the laborious and costly process of selective breeding for plants or animals was no longer necessary. Moreover, you weren't limited to genetic changes from the same or closely related species. For example, you could direct a batch of bacteria to produce any kind of protein you wanted it to produce, as long as you could find a gene in any species that produced that protein and then splice the gene into the bacterial DNA. Because bacteria reproduce every twenty minutes, you would shortly have a huge vat full of these bacteria creating that protein.

That produced tangible results quickly. In 1976, Herb Boyer founded the company Genentech. In 1978 the company engineered the gene for human insulin into *E. coli* becoming the first FDA-approved genetically engineered drug for human use in 1982. Prior to that, insulin was extracted slowly and expensively from the pancreas of pigs and cows. That was only the start. A revolution had begun.[234]

From bacteria to humans.

The revolution took off in plants. Genetically modified plants have been used for food for decades beginning with the Flavr Savr tomato in 1994. It was transgenically engineered to delay ripening after picking. Subsequently, genetically engineered corn, canola, soybeans, and many other crops have become common in our food supply. Engineered traits include resistance to pathogens and herbicides, improved nutritional value, improved appearance, and many other phenotypic characteristics.

It is one thing to insert genes into a plant or a bacterium, but what about inserting genes into a human? The jump from one to the other wasn't without many intervening steps.

In 1974, Rudolf Jaenisch performed genetic engineering on a mouse.[235] A fascinating experiment was done shortly thereafter on a rabbit. DNA from a jellyfish made the rabbit glow in the dark.[236] The point was not to create a pet that was easy to spot at night but to demonstrate how readily a variety of genes could be inserted into mammals for medical purposes—including genes from humans. For example, a human gene that produces a substance that prevents blood clots (antithrombin) was introduced into goats. These transgenic goats produce milk containing antithrombin that can be easily extracted for medical use in humans. The FDA has approved these so-called pharm animals.

Of course, this has not been without public controversy and political debate. The first Asilomar Conference in 1975 raised the potential dangers of recombinant DNA and genetic engineering.[237] Although voluntary and useful safety recommendations came out of the conference, it also illustrated how difficult it is to control this area, given the swift evolution in related technologies and the cumbersome and varied political environments attempting to control them.[238] With regard to genetically modified organisms (GMOs), particularly those modified for food, concerns have been raised ranging from potential harm to consumers to environmental impact. This has resulted in demands that these products be labeled and even banned. Although some countries require labeling, the United States and Canada have resisted this requirement because there is no widely accepted scientific evidence that genetically modified foods have any more risk than plants or animals modified by selective breeding. The public pressure continues, however, and as of this writing, various labeling bills are still in process through both houses of the United States Congress.

The controversy, however, has had an impact on the approval of genetically modified food animals as opposed to plants. It wasn't until November 2015 that the FDA approved the first transgenic animal to be sold as a food product. This is an Atlantic salmon that contains genes from two other species causing faster and larger growth. The FDA, in an abundance of caution, has placed restrictions on the production of these salmon so that they will be isolated in special breeding tanks located outside of the United States and in a manner that would prevent escape into the wild. However, the salmon will not require special labeling when sold, at least as of this writing.

GMHs

If public concern and governmental caution are high regarding GMOs, there is an especially high concern for one particular subcategory of GMO: GMHs or genetically modified humans.

Thanks to the Human Genome Project, we now know the full DNA sequence of *Homo sapiens*. This massive collaboration involving about twenty research centers around the world was begun in 1990 and completed in 2003. The human genome is a sequence of about three billion nucleotide base pairs. This creates an interesting twist. Although there are more than seven billion humans alive today, no two humans have the same complete nucleotide sequence, including so-called identical twins. So whose genome did the Human Genome Project sequence? It turns out that less than 0.1 percent of our genome accounts for all of the variation in the human population! Therefore it is a moot point whose genome it is; it doesn't matter in the big picture. It does matter, however, when we genetically modify an individual human. What took twenty research centers thirteen years to accomplish originally can now be done in a few hours. Any individual's genome can be sequenced quickly. This makes GMHs a practical consideration.

Genetically modifying humans is currently driven by the desire to cure genetic disease. To consider gene therapy for genetic diseases, we first need to know which genes are causing disease or abnormalities. In addition, we now know from chapter 5 that it is not simply a matter of knowing which gene is involved, but how that gene is expressed through its regulators—epigenetics. It is turning out to be very complicated.

One powerful tool is an online database that catalogs human diseases with a known genetic basis called Online Mendelian Inheritance in Man (OMIM).[239] It was originally created and maintained by Dr. Victor McKusick of the Johns Hopkins School of Medicine. Updated daily from contributors around the world, it provides detailed genetic information, phenotypic descriptions of the diseases, references, and a wealth of other information for use by researchers anywhere. As of July 2015, the database contained descriptions of 5,523 disease phenotypes with a known genetic basis and described the exact mutations of 3,417 genes causing disorders.

Dr. McKusick's pioneering work enabled the first necessary step required for gene therapy. Thanks to OMIM, we now know thousands of genes that cause thousands of clinical disorders.

There are two general types of defective genes:

1. Those with mutations that prevent the production of a necessary protein.
2. Those with mutations that cause the production of a harmful protein.

For the first type of defective gene, gene therapy involves replacing that gene with a healthy one or adding an additional healthy gene. For the second type, gene therapy needs to either replace that faulty gene with a healthy one, or disable the defective gene in some way.

Techniques for identifying and isolating defective genes for recombinant DNA have now matured to the point where this can be accomplished with most known defective genes. After the gene is isolated, a sufficient amount of recombinant DNA must be produced in order to populate the host organism. To that end, in 1983 biochemist Kary Mullis developed a mechanism called polymerase chain reaction (PCR) that can quickly replicate millions of copies of any DNA sequence. This earned Mullis a Nobel Prize.[240]

We then need to be able to deliver the healthy gene or gene segment to the appropriate cells of the patient. Inserting the healthy recombinant DNA into a virus (called a vector) is one means of doing this. Viruses know how to invade cells. If viruses are used as vectors, they must be altered so that they can still infect the target cells but not introduce viral disease in the treated patient. For example, if the defective gene affects the lung tissue, the viral vector chosen normally causes respiratory infection. In its altered recombinant state, when the virus is delivering the healthy gene to the lungs, the virus still can penetrate the respiratory cells but not cause respiratory disease. Alternatively, the recombinant DNA vector can be delivered to the lungs through inhalation therapy. For diseases affecting the blood or bone marrow, the blood cells or bone marrow cells can be removed from the body, grown in a culture medium outside the body, combined with the recombinant DNA in the culture medium, and then injected back into the blood stream. Another example is an early attempt to genetically treat a disease of the retina of the eye by injecting the viral vector directly into the fluid of the eye adjacent to the retina.[241] The point is, there needs to be some way to get the recombinant DNA into the right cells in the body in order to be effective. That is not easy.

Early gene therapy experiments.

The first experiment in human gene therapy was performed on a patient with a rare disease called severe combined immunodeficiency syndrome.[242] Children with this disease die from the inability to fend off infections. The cause is a genetic defect in which a normal enzyme needed for immunity is not produced by the white blood cells. The exact gene that produces this enzyme is known.

In 1990, a group of researchers at the National Institutes of Health took blood from a patient with this disease and extracted the white blood cells that normally should have been producing this enzyme. Using a viral-based vector, they produced recombinant DNA with a healthy copy of the gene, incubated it with the white blood cells of the patient, and injected them back into the patient. It worked to a limited degree. The patient definitely benefitted from the therapy, but it was short-lived, requiring periodic additional gene therapy for the rest of her life as well as supplemental therapy with a medication providing the enzyme itself.

Although this first gene therapy patient grew to adulthood and lives a relatively normal life, her case illustrated some of the difficulties in bringing gene therapy into the mainstream. The recombinant DNA-treated white blood cells did produce the normal enzyme, but they did not pass on this ability to subsequent white blood cells produced in her body. That is because white blood cells are produced from stem cells in the bone marrow that were not genetically engineered. For that reason she needs repeated gene therapy treatments as well as other treatments.

Another genetically based immunodeficiency disorder has a completely different gene mutation as its cause. Ten patients with this condition were treated with recombinant DNA using a different virus vector and a healthy copy of the gene. This time, the treatment involved replacing bone marrow with new cells treated with the recombinant DNA. The resulting healthy bone marrow cells repeatedly produced new healthy white blood cells, and nine of the 10 patients were cured. However, three years later, two of the patients developed leukemia. It was determined that the virus vector had the unanticipated consequence of altering another gene that led to the cancer, known as off-target mutation.[243] It took another 10 years of research to determine the genetic mechanism of the off-target leukemia side effect. A different viral vector combined with low-dose chemotherapy has eliminated this problem and patients are again being cured.[244] However, off-target mutations continue to plague gene therapy. They call into question the precision of genetic engineering.

One of the worst early gene therapy disasters occurred at the University of Pennsylvania in 1999. An eighteen-year-old patient with a genetically caused liver disorder was treated with recombinant DNA containing the healthy gene. The patient suffered a massive allergic reaction to the virus vector and died. Federal investigators found serious lapses in both scientific and procedural aspects of this case followed by calls for stronger regulation.[245] The impact of this case still lingers today.

A tale of two genes.

Let's revisit two diseases discussed in the chapter on speciation and natural selection: cystic fibrosis and sickle cell anemia. Each of these relatively common genetic disorders in humans involves mutations in a single gene. The most common gene mutation in cystic fibrosis involves only three nucleotides. Sickle cell anemia is caused by a mutation affecting only a single nucleotide. In both cases, the mutation results in the failure to produce a normal form of a necessary protein. Since every cell contains two copies of each gene, in those people with one normal copy and one diseased copy (the carrier state), the normal copy produces enough of the necessary protein to prevent symptoms. However, as we learned from Mendel's peas, on average one-quarter of children from two carrier parents would inherit two copies of the mutant gene and thus have the disease.

Since the genetic abnormality in each of these diseases involves a single gene, and in fact only one or three nucleotides in that gene, one would think it should be relatively easy to cure them with genetic engineering. It is not.

Cystic fibrosis appears in childhood and affects both the lungs and the pancreas. Although various treatments usually prolong the lives of patients into adulthood, the treatments are continuous and expensive, and the patients often have multiple complications sometimes shortening their lives. The problem with gene therapy is in delivering the recombinant DNA to both the lung and pancreas tissue affected. Most of the efforts are aimed at the lungs because that is where the most serious disease complications arise.

Since 1989, there have been over one thousand gene therapy clinical trials for cystic fibrosis using a variety of vectors and techniques for delivering the healthy gene to the lungs. The results have been mostly disappointing but with some small successes. Some of the techniques do demonstrate improvements in lung function, but they are usually short-lived requiring repeated treatment. At this point in time, there is no gene therapy cure, or even an agreed-upon long-term treatment using gene therapy for cystic fibrosis. The research is continuing, and hopefully the goal will be achieved within the next decade.[246]

A similar problem has occurred in trying to cure sickle cell anemia—delivering the healthy gene segment to the right cells. The protein involved is hemoglobin—contained in red blood cells. Diseased patients contain abnormal hemoglobin that causes the red blood cells to "sickle" when circulating which leads to painful clogging of blood vessels and serious complications. Since red blood cells are the only cells in the body that do not have nuclei, they do not contain genes. So the therapy must be aimed at the bone marrow precursor cells. If one simply removes a sample of these precursor bone marrow cells, genetically engineers them, and infuses them back into the patient, these cells do function normally. The problem is that since the bone marrow still contains the old diseased cells that are still producing abnormal red blood cells, the patient still has sickling. Finally, after decades of attempts, we have the early report of a sickle cell anemia cure using a very complicated combination of genetic engineering and chemotherapy.[247]

This has not been easy.

It is clear that the early bloom is off the rose for gene therapy. Progress has lagged since the excitement and even hype that characterized the scientific community ten to twenty years ago. There have been many disappointments and even disasters. Nonetheless, glimmers of hope and some successes shine through. There are even some cures or at least long-term improvements in some experiments in patients.[248 249] These successes are paving the way for much greater success in the future. It is only a matter of time.

The first (and as of this writing, only) commercially available gene therapy is a drug called Glybera, approved only in Europe for the treatment of a very rare genetic disorder causing pancreatitis. It involves replacing a defective gene using a viral vector. The drug is injected directly into muscles, which normally produce the enzyme. It is very expensive and only partially effective.[250] Near the end of 2016, dramatically successful clinical trials were completed on another gene therapy drug, nusinersen, for the treatment of an uncommon neuromuscular disease in children called spinal muscular atrophy.[251] FDA approval was expected shortly.

Another very promising use of genetic engineering is an approach to cancer therapy called CAR-T (Chimeric Antigen Receptor T cells). T cells are blood cells that are part of the body's normal immune system to fight infections. In CAR-T therapy, a patient's T cells are collected and then genetically altered to recognize the patient's specific cancer cells. The genetically altered T cells are grown outside the body until they multiply into billions of cells that are then injected back into the cancer patient. These cells circulate, and when they find the specific cancer cell, they destroy it. Early clinical trials of CAR-T therapy are very encouraging.[252 253]

Another novel approach to cancer using genetic engineering is being studied to fight prostate cancer. Using an adenovirus vector, the prostate cancer tumor is injected with this vector to genetically alter its cells to contain a herpes-virus gene. An antiherpes drug is then administered that targets and kills these genetically altered prostate cancer cells.[254] This therapy is also in the clinical trial phase and is showing promising results.

We're at the point where our genetic engineering is getting easier, but the clinical implementation of it remains difficult.

Germline genetic engineering

So far, all of the genetic therapy trials in humans have involved somatic cell genes (including cancer cells derived from somatic cells) of individual patients—not germline cells. Remember from chapter 5 that there are two categories of cells in the human body: most of our cells are somatic cells that form all of the tissues and organs in the body; some of our cells are the specialized reproductive cells in the testis or ovary, called the germline cells. Alterations in the genes of somatic cells affect only the cells that are altered and possibly any new cells produced by mitosis from these cells. Gene therapy aimed at these

International Survey

A survey of thirty-nine countries performed in 2014 indicated that twenty-five of them had legislation banning germline therapy, four of them had "guidelines" banning it, nine had "ambiguous" policies, and one had "restrictive" policies. Enforcement was considered potentially weak in many of these countries.[255-257]

cells will affect only the individual patients given the gene therapy, not their offspring.

Gene therapy could be applied to germline cells as well. Germline cells include sperm, eggs, embryos (zygotes), and embryonic stem cells—the cells that produce sperm and eggs. Alterations of genes in these cells would be passed on to future generations. That makes safe engineering of somatic cells imperative before we can ethically and legally move on to germline cells. Today it is considered unethical to attempt to perform gene therapy on human germline cells. (See appendix 5 for a discussion of a form of germline gene therapy that was attempted for a brief period in the United States.)

In fact, not just unethical but often illegal (see box).

Undoubtedly, the legality of germline genetic engineering in at least certain countries is only a matter of time. In February 2016, Great Britain approved germline modification of *in vitro* embryos on the condition that they be used only for research purposes and not implanted. It is a start. Great Britain had already approved mitochondrial gene replacement therapy in 2015 (see appendix 5). In 2016, an Institute of Medicine committee recommended to the FDA that "cautious" experimentation with mitochondrial gene replacement research be allowed in the United States, but that only male embryos be allowed to be implanted in women with the intent to proceed to live birth.[258] In that way, this form of germline genetic modification could not be passed on to any offspring of such a live birth (because mitochondria are inherited only from the mother).

There are many potential benefits of germline therapy. For instance, imagine that you undergo genetic testing and are told that you have the gene for Huntington's disease. People with this gene are usually normal until middle age. Then they start to develop neurological problems that evolve into grotesque involuntary movements, dementia, and usually premature death. Because it is an autosomal dominant gene, you will surely get the disease. Each of your offspring will have a 50% probability of getting your abnormal gene.

What are the options for people who learn from genetic screening that they have the gene for Huntington's or some other horrible disease? They could possibly attempt somatic genetic therapy on themselves, if such an approved clinical treatment or clinical trial is available. Over time, such therapy will become available for increasing numbers of disorders. This does not, however, prevent passing on the abnormal gene to their children.

A second option is becoming increasingly available. It is called preimplantation genetic diagnosis (PGD), and is currently used as a part of *in vitro* fertilization (IVF). At an early stage during IVF, when the embryo has only a few cells that have not yet begun differentiating into the various tissues, it is possible to remove the DNA from one of these cells and do a complete genomic analysis. This does not interfere with the subsequent health of that implanted embryo. If a serious genetic defect is discovered in this analysis, that embryo would be discarded. In the case of Huntington's disease, for example, on average only one-half of the embryos created from a parent with the abnormal gene would have that gene, and one could simply select an embryo with a healthy version of the gene for IVF. It is only recently that the cost, speed, and accuracy of full genome analysis, coupled with our increasing knowledge of disease genetics, have made PGD feasible. One still needs to do the IVF procedures, and PGD raises a number of ethical issues as well if the technique is used for less serious conditions or even cosmetic traits. Also, although it does prevent passing on the abnormal gene to the children, it does not prevent the disease in the parent.

Recommendation
For an excellent discussion of the science, medical, technical, and ethical issues surrounding PGD, I recommend Henry Greely's book The End of Sex and the Future of Human Reproduction.[259]

The best solution—if it were technically possible, medically approved and ethically accepted—would be to try to eliminate that gene in both somatic cells and germline cells so that they couldn't be passed on to future generations. Why would we want to correct a terrible genetic disorder in a person only to have it passed on to his progeny? We certainly aren't close to being there yet, but sooner or later that is going to change. I can't tell you when it will change, but it may not be as far into the future as you might think.

The next steps

Enter CRISPR, or Clustered Regularly Interspaced Short Palindromic Repeats. Quite a mouthful!

For decades, researchers have observed a strange sequence of DNA nucleotides at the end of some bacterial genes. This is what the term CRISPR describes. At first they thought this was just junk DNA. However, in 2005 it was shown that a part of this odd-looking sequence actually corresponded to a piece of the DNA of certain viruses that can infect bacteria. Subsequent research showed that this was not junk, but rather an effective defense system of the bacteria against these viruses. The CRISPR part of the genes actually create RNA molecules that, in conjunction with a protein called Cas9, cut up the invading virus DNA and destroy it. (CRISPR is sometimes referred to as CRISPR/Cas9). Cas9 is an enzyme called a nuclease that actually does the cutting. The CRISPR part of the system tells the Cas9 part where to do the cutting.

Eureka! It became apparent that if you could change the CRISPR sequence with some DNA editing, you would have a general-purpose tool that could cut any DNA anywhere by using the modified CRISPR to direct the Cas9 enzyme. That's exactly what has now happened.[260] All of the previous discussion so far about gene therapy in this chapter relied on the fact that the researcher could identify a good gene somewhere, cut it out and splice it into recombinant DNA with a vector, and then insert it into cells with the bad gene. What I haven't said is that the scissors currently used to "cut out" the good gene and splice it into the recombinant DNA are not simple tools. They are all labor intensive, take months to years to make for any given target gene, are expensive, and are subject to error.

CRISPR/Cas9 changes all of that. It can be set up cheaply in just a few weeks for virtually any gene. In fact, a do-it-yourself CRISPR kit with enough material to perform five gene editing experiments is available online for $150. Suddenly, a tool had been found that in less than a year's time since its discovery was being used "to delete, add, activate, or suppress targeted genes in human cells, mice, rats, zebrafish, bacteria, fruit flies, yeast, nematodes, and crops, demonstrating broad utility for the technique."[261] Some have labeled this the Swiss army knife for genetic engineering and gene therapy.[262]

Because many genetic disorders are caused by multiple gene mutations, CRISPR/Cas9 can be set up to deal with multiple genes simultaneously. It can be used in somatic cells, stem cells, and germline cells. Animal models of human diseases can be created in a matter of weeks rather than the months or years required with previous methods. New companies are sprouting up to market commercial uses of CRISPR. Patent wars have already begun over its ownership.[263] "The only limitation today is people's ability to think of creative ways to harness CRISPR."[264] CRISPR was named *Science Magazine*'s "2015 Breakthrough of the Year." See box.

One example of CRISPR use is being experimented with in the fight against malaria.[266] CRISPR is already being used in China to produce goats that have larger muscles and longer hair.[267] This creates more profit for goat herders, who can sell more meat for markets and wool for cashmere clothing at lower production costs. In July 2016, a Chinese investigator announced the first human clinical trial using CRISPR for a cancer treatment in patients (not a germline treatment).[268] In December 2015, the International Summit on Human Gene Editing, sponsored by scientific organizations in the United States, China, and United Kingdom, was held in Washington, DC, to discuss the overall science and ethics of this technique. Their concluding statement reaffirmed that germline genetic engineering in humans should be restricted to research on embryos and gametes only for *in vitro* settings, but they did condone continuing research with the view to revisit this recommendation in the future. It is CRISPR/Cas9 that was approved for research on *in vitro* human embryos in Great Britain.

Nature Magazine on CRISPR, July 2016:
"Today, CRISPR is a household name for molecular biologists around the world. Researchers have eagerly co-opted the system to insert or delete DNA sequences in genomes across all kingdoms of life. CRISPR is being used to generate a new breed of genetically modified crops and may one day treat human genetic diseases ... principal investigators involved in the seminal work have become scientific celebrities: they are profiled in major newspapers, star in documentaries and are rumoured to be contenders for a Nobel prize."[265]

We've seen this movie before.

There is an all-too-familiar pattern to scientific "breakthroughs": discovery, hype, and then disappointment. We're at the hype phase for CRISPR. There are many caution flags being waved. Until we get to definitive clinical trials demonstrating efficacy and safety in human gene therapy, we will not know for sure how valuable this tool really is.

One of the cautions raised is that CRISPR may have greater potential than other techniques for off-target mutations.[269] That is, it could cause unintended deleterious mutations in genes other than the targeted gene—similar to the problem discussed earlier in the "cured" immunodeficiency patients that later developed leukemia. Other cautions have been raised as well.[270]

In a study done at a university in China, a group of researchers experimented with using CRISPR to genetically modify human embryos.[271] They were using nonviable embryos supplied by a fertility clinic and had no intention to use those embryos for any purpose other than laboratory research. Nonetheless, the paper was significant for two reasons.

1. It demonstrated that, indeed, the CRISPR/Cas-9-mediated gene therapy caused off-target mutations and other problems. They stated, "Our work highlights the pressing need to further improve the fidelity and specificity of the CRISPR/Cas9 platform."
2. It set off a firestorm of debate on the ethics of doing genetic engineering experiments on human embryos because these are germline cells whose changes would be inheritable if ever implanted in a human and resulted in a live birth.[272] An entertaining discussion of this controversy aired on NPR's Radiolab.[273] It is noteworthy that the conclusion of the International Summit on Human Gene Editing, cited above, would allow this research.

In September 2016, another NPR radio segment reported on the use of CRISPR to modify healthy, viable human embryos by a researcher at the Karolinska Institutet in Sweden.[274] Again, there was no intent on using these modified embryos for implantation, but it demonstrates the inexorable march toward such a possibility.

Some countries are clearly more tolerant of gene editing and manipulation than others. For example, cloning is a form of genome manipulation in which the entire genome of one individual is implanted into an enucleated egg (i.e., all of its DNA has been removed) of another individual and then implanted into a uterus to allow a normal pregnancy. The resulting newborn has the identical genetic makeup of the donor genome individual. There is a company in China, called Boyalife, which plans to clone one million cows per year by 2020 by cloning particularly desirable cows. Its South Korean partner, Sooam, now clones dead pet dogs to bring a loved pet "back to life" for some customers, and it is planning to clone an extinct woolly mammoth.

Our understanding of genetics, DNA, cloning, and genetic engineering is growing at an exponential rate. What was science fiction a decade ago is now reality. We are now at the point where we don't even need to cut out a gene and splice it into a vector to create a new gene sequence. As long as we know the DNA sequence of a gene of interest, we can now make it from scratch in a laboratory and insert it into a vector. It's like a DNA typewriter. We can even create an entire genome this way and implant it into a bacterial cell that has had its natural DNA removed from its nucleus, and that bacterial cell will function and replicate with the new DNA.[275] We are very close to synthesizing the entire genome of a yeast species,[276] which would be the first Eukaryote (the domain that includes all plants and animals) so created. Synthetic, parentless organisms! You can order any gene you want to be synthesized from commercial companies if you know its code. There was a meeting at Harvard University in 2016 of 150 scientists, lawyers, and ethicists to discuss the possibility of synthesizing from scratch a total human genome within ten years.[277] Such a genome could theoretically be customized in any conceivable manner. It could then be inserted into an embryo to replace the natural DNA and develop into a parentless human. Is this our path to *Homo nouveau*?

What will be the first germline experiments?

All of the scenarios that could lead to the creation of *Homo nouveau* would require the alteration of germline cells. We are clearly years or even decades away from doing germline gene therapy in humans (with the exception discussed in appendix 5 regarding mitochondrial germline therapy). But when (not if) we begin doing it, we can start thinking about if and how it will be used to create *Homo nouveau,* whether using natural or synthetic genes.

Germline therapy in humans has already begun with mitochondrial germline therapy. The next least controversial area could be spermatic stem cell therapy to cure infertility in some males. This does not involve altering embryos—a politically sensitive issue. It only involves altering and examining the ubiquitous sperm produced by these stem cells prior to fertilization.

Next, germline therapy will probably be used for the elimination of known disease-causing genes, particularly those caused by a single defective gene. Only if proven safe and effective on animal models and in clinical trials would it become available to the general public. Over time, assuming success in these less complicated forms of germline therapy, it will become politically acceptable to move on to more complex areas.

The next most popular use of germline gene therapy may be to produce "designer" babies. Even though that's what most people fear will happen, we already are seeing the sale of designer "micropigs" as pets in China.[278] In humans, we have designer babies to a small degree, with *in vitro* fertilization where the sex of the implanted embryo can be selected. It is notable that such sex selection was banned by the Oviedo Convention in 1997, which simply illustrates how mores and attitudes can change once a technology becomes well established. Height, weight, hair and eye color, presence or absence of freckles, and many other phenotypic characteristics will be potentially designable. That depends on whether or not the genetics are clearly understood, and whether the desired genes are available in the parents or elsewhere. If the parents don't have them, one can assume there will be a market for donor genes in the same way there is for donor sperm.

What else might be possible? What about more complex phenotypes like intelligence, mathematical ability, athletic skill, or beauty? That will depend on our state of knowledge of the genes associated with those traits and how those genes are expressed. The genetic determination of some traits will be unknown, too complicated to genetically engineer or even too risky to genetically engineer.

A likely popular genetic engineering target.

What about the complex problem of aging? Why do we grow old? Why can't we keep rejuvenating our skin or our brains? Depending on the tissue, cells are constantly dying and being replaced with new cells by mitosis. That is particularly true in the blood, skin and colon. It is not true of the neurons in the cerebral cortex, which generally live our entire lifetime and are not replaced when damaged. Nonetheless, even tissues that are continually renewing themselves, like skin, age.

Diet is one factor that definitely affects aging. Mild to moderate caloric restriction has been shown to increase lifespan in most animals studied.[279] Some of the molecular pathways mediating this effect are known.[280] Could we engineer genes to do what diet restriction does?

The research on genes and aging is voluminous. Clearly, something happens to some parts of your DNA over time as the body ages. Remember, as new cells are created to replace old cells in your body, your DNA replicates itself during mitosis. This replication process is not exact so the more replications that occur, the more that the DNA in those cells diverges from what it looked like at birth. Some of those changes are related to aging.

One area regarding aging that shows promise for possible gene therapy relates to a part of the DNA called telomeres. These are short, repetitive DNA sequences on the ends of each chromosome. During each cell division, one or more telomere is deleted as though a ticket had to be taken for each mitosis. Eventually, there is little or no room for shortening, which is related to the senescence or aging of the cell.[281] In 2015, the first person reported to have genetic engineering to increase telomere lengthening was the CEO, named Elizabeth Parrish, of a US-based company. The company, BioViva, is dedicated to anti-aging technologies. Since this type of therapy is not approved in the US, she had the procedure performed in Columbia. The procedure involved genetically engineering her white blood cells and re-injecting them back into her. The company reports that her current circulating white blood cells, as measured by telomere length, are now 20 years younger.[282] These results have not been verified in the peer-reviewed literature and should be treated skeptically. But the very fact that there are companies dedicated to anti-aging technologies and experiments are already underway indicates the great interest in this area. See box for other aging factors that could be subject to genetic engineering.

An online database called GenAge tracks research on genes that may be related to aging. As of January 2015, there were 298 such genes under study.[284]

These genes affect a wide range of cellular and metabolic processes that may be related to aging and age-related diseases.[285] Additionally, SENS (Strategies for Engineered Negligible Senescence) Research Institute is one organization devoted to studying diseases of aging. Their goal is to find ways to repair the molecular and cellular damages of aging and to rejuvenate tissues. This includes ways to genetically engineer mitochondrial genes.[286]

Aubrey de Grey, a gerontologist and chief science officer of SENS Research Institute, has popularized the term Methuselarity to indicate the point at which our ability to reduce the effects of aging outruns the aging process.[287] This, which could lead to almost unlimited life spans will come about by progressive and incremental improvements in our ability to restore damaged cells to their predamaged state—so-called regenerative medicine. Much of this, but not all, will occur through genetic engineering.

Other genetic aging factors

Repeated mitochondrial replication leads to progressive damage of mitochondrial DNA, which may lead to cell senescence. Epigenetic processes probably play a major role in the processes of cellular senescence. Recent studies in mice have shown that these epigenetic processes can be engineered to be reversible; in essence making old cells become younger in certain tissues of these mice.[283]

It probably will be a longer time before gene therapy to slow aging will become a reality compared to gene therapy for many diseases. One can even question whether achieving the Methuselarity, or even coming close to it, is a good thing. Its impact on our already burdensome population growth would be negative. What about the financial impact? What if the results are not uniform on all tissues of the body? It seems there might be a lot of new problems to deal with. Regardless, any success in this area is surely destined to be one of the most popular genetic engineering procedures.

The Big Question

The question for us is whether this could lead to *Homo nouveau*. In my view, not likely. Although we'll have a lot more younger-looking and healthy older *Homo sapiens*, it won't likely lead to cladogenesis. Even if it increases the number of fertile years, I see no reason why interbreeding patterns would change because of that.

One could argue that delaying or preventing death slows evolution. Daniel E. Koshland Jr., the former editor of *Science* magazine, states that death plays two very important roles in sustaining a species.[288] Without aging-related death, we wouldn't need to reproduce very often. That would deprive us of our most important mechanism for survival. Reproduction provides the best mechanism for regeneration of tissues. By creating a new, fresh copy of organisms from germline cells, all organs and tissues are regenerated. As discussed in chapter 5, the DNA slate with all of its accumulated changes during a single lifetime is wiped clean, with the possible exception of some epigenetic changes. Second, reproduction by meiosis provides the opportunity for genetic recombination, which is our best mechanism for adaptation. Without death and birth, we wouldn't have natural selection. Therefore being good at delaying or preventing death may be good for an individual, but it is bad for the species.

In any case, there will be great debate over the ethics of performing gene therapy other than to prevent serious morbidity or disability. Some even argue against it for any reason. The arguments range from religious reasons (humans should not "play God") to implications that the lives of the disabled are not of value, to ethical concerns about genetic engineering becoming tools for eugenics with all of its negative consequences. In my view, it will be similar to the debate that we had over *in vitro* fertilization. We'll get over it once we know how to do it safely.

So far, nothing in this discussion seems likely to lead to speciation of *Homo nouveau*. All of the engineering that has been discussed so far would produce cisgenic humans. There is no obvious reason why the adults from any of these GMHs wouldn't freely interbreed with the general population and produce viable offspring.

How could we genetically engineer Homo nouveau?

Could we genetically engineer a trait that doesn't occur naturally in some existing species? Perhaps that might more likely lead to *Homo nouveau*. For example, could we genetically engineer a biological brain wave detector? No, we couldn't. There is no way we could devise the nucleotide sequence that would do that, at least not with our current knowledge. We can only use (or copy) genes that already exist in some species. On the other hand, could we genetically engineer a human to have retinal receptors for ultraviolet light? Yes, we probably could because genes for such receptors exist in birds and other species; that would produce a transgenic human. It is certainly conceivable.

Would such an experiment be allowed? Maybe. We've allowed an organ transplant from a baboon or a chimpanzee (xenotransplantation), but only in the most desperate clinical situation where there was no other treatment or suitable human organ donor available. Transgenic gene therapy seems less outrageous than that. There is a great deal of work going on at present to genetically modify pig organs to make them safe for human transplantation.[289] There certainly would be an ethical debate about performing transgenic gene therapy directly in humans for reasons other than saving lives. Again, given enough time, I think we'll allow it. If someone wants to see ultraviolet light, why not let him? I can think of lots of useful roles for a person like that. We allow people to have facelifts, and cosmetic surgery is a legitimate specialty. Why not genetic surgeons?

The real debate would be between doing such transgenic gene therapy on germline cells versus somatic cells. It is one thing to allow a person to see ultraviolet light if he wants to. It is quite another to allow him to pass on that trait to some or all of his offspring over generations. Would he be required to disclose this to his wife to be? I doubt that will be an issue. We're not required today to disclose everything in our genetic makeup to anyone, and there could be a lot of really bad stuff there.

Once we're as good at doing germline therapy as we are any other medical procedure, it will be allowed. I'm confident there will be gene therapy breakthroughs beyond CRISPR that will make all of this as routine as cosmetic surgery. To that end, another CRISPR system that uses a different "cutting" enzyme than Cas9, called Cpf1, has already been reported.[290] Other techniques have also been reported that reduce the risk of off-target CRISPR errors.[291] One such technique is borrowed from the never-ending war of escalation between viruses and bacteria. The CRISPR/Cas9 system is an evolved bacterial defense against viruses. In response, some viruses have evolved an antidote to Cas9—a protein that turns off the Cas9 ability to cut DNA. Research is ongoing to potentially use this viral-produced protein in humans to reverse any off-target side effects of CRISPR-based genetic engineering.[292] Recently, a new CRISPR system was reported that would allow the alteration of a single nucleotide rather than entire genes.[293] The rate of progress is astounding. Yes, there will always be some risks, but that's why we require informed consent for elective procedures.

By now, it appears that the views of Jean-Baptiste Lamarck have been resurrected. Because of germline therapy, traits acquired after birth *can* be passed on to offspring. We're also getting a little closer to thinking about *Homo nouveau*. It seems likely that somewhere and sometime, someone will produce humans that, for one reason or another, will be reproductively isolated from the rest. It is just as likely that it will occur by accident as by design because of off-target mutations.

Here's one possible scenario. Suppose there is a popular genetic alteration—such as the prevention of Alzheimer's disease—and thousands of these are done every year around the world. Further suppose that this alteration was associated with an off-target mutation that caused the offspring of these people to have a high spontaneous miscarriage rate when mating with anyone other than those with the same targeted alteration. It could take several generations before anyone figured out what was going on. By then, we could have the start of a separately evolving metapopulation (i.e., *Homo nouveau*). The separate lineage could become reinforced if, say, further mutations occurred later in that lineage that made it less likely that any surviving hybrids could have viable children. This scenario could occur regardless of whether the alteration was cisgenic or transgenic. Figure 21 is the de Queiroz diagram of this situation.

FIGURE 21

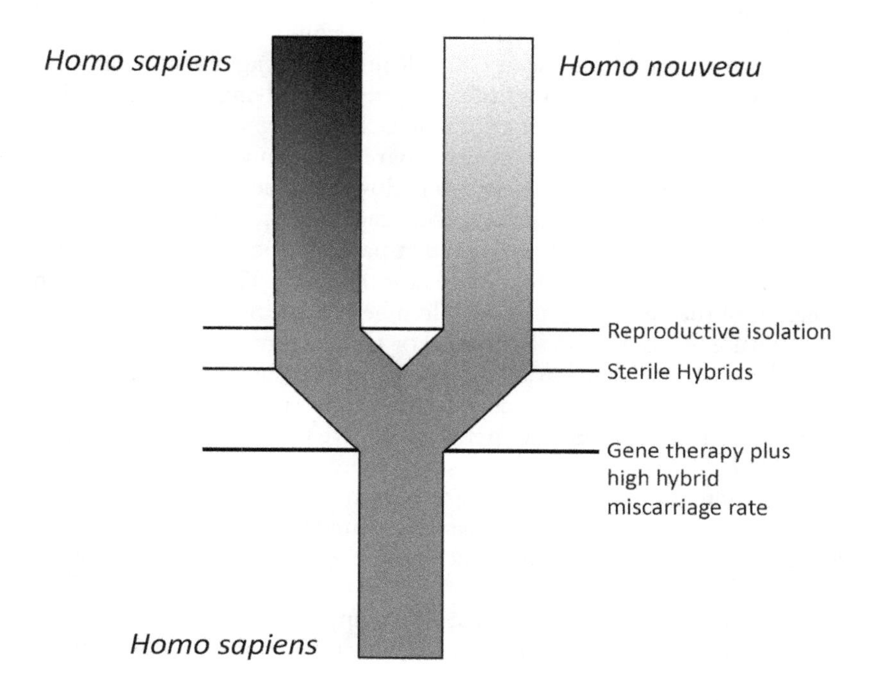

Suppose now that after a few generations, some astute clinical researcher figures out what is happening. By then there could be 100,000 or more of these *Homo nouveau* that learn they can only have children by mating with another *Homo nouveau*. How many would have gene therapy to try to have this condition reversed? It depends on how important their new genetic makeup is to them. Maybe *Homo nouveau* would be short-lived. On the other hand, maybe they would prefer their new genetic makeup, and we could add another SC to figure 21 for behavioral isolation.

That raises another possibility that Lee Silver described in his book *Remaking Eden*.[294] He envisions a future society practicing an extreme form of behavioral isolation based on genetic engineering. In this society, only a small portion of the population, which he calls the GenRich, have the financial means to genetically enhance their children. Using a process he calls reprogenetics, the GenRich have used genetic engineering techniques developed over decades that allow them to optimize a wide variety of human traits (including intelligence, athletic skill, physical appearance, creativity and many others) to put the GenRich in a controlling position throughout society.

Over time, the wealth and cultural disparity between this GenRich minority population and the remaining "naturals" have become so great that there is little financial mobility or voluntary interbreeding between the two groups. Such a scenario could lead to cladogenesis by some chance genetic development of a postzygotic reproductive barrier.

My sense is that in the early years of this process, there would be too much interbreeding of the early GenRich with the "naturals" to allow them the time to evolve separately, but this is certainly one possibility.

There is a variation to the use of CRISPR in genetic engineering that is potentially even more likely to lead to a separately evolving metapopulation. Normally, an embryo receives one copy of each gene from each parent. Thus a given gene variant inherited from either parent has only a 50 percent chance of being passed on to the next generation. However, one can now use a genetic tool, called gene drive that causes a specific gene inherited from one parent to alter the corresponding gene from the other parent to be identical to it. This in essence makes the offspring homozygous for that trait. That would mean that 100 percent of the next generation would inherit the trait.* Gene drive therapy is being considered as a means to control or eliminate mosquito populations carrying diseases such as malaria or Zika. Currently gene drive is still in the experimental stages, and there is considerable concern that this technology could inadvertently introduce unintended negative traits into the environment that would be difficult to control. However, there is no technical reason why gene drive couldn't be used in the future in humans. Although gene drive would not fundamentally change the issues related to speciation of *Homo nouveau*, it speeds up the rate at which such an event could occur through genetic engineering.

I'm sure others can and have dreamed up many other scenarios that could result from human genetic engineering. Nicholas Wade, in his book *Before the Dawn*, envisions creating an extra chromosome in human germline cells that contains an entire suite of genes that correct various genetic defects and even add some enhancements.[295] Each individual would then have a custom-made extra chromosome. The potential scenarios seem endless.

How likely is any of this?

Many very informed people are quite alarmed about the possibilities. Bill Joy, the cofounder of Sun Microsystems, wrote in his famous essay in *Wired Magazine*: "If, for example, we were to reengineer ourselves into several separate and unequal species using the power of genetic engineering, then we would threaten the notion of equality that is the very cornerstone of our democracy."[296] He goes on to say, "I think it is no exaggeration to say we are on the cusp of the further perfection of extreme evil, an evil whose possibility spreads well beyond that which weapons of mass destruction bequeathed to the nation-states, on to a surprising and terrible empowerment of extreme individuals." He calls this "knowledge-enabled mass destruction."

Joy's solution is voluntary relinquishment—that is, agree not to do it. He cites our voluntary prohibition of biological weapons as a precedent. I don't foresee that happening. Unlike biological weapons, human germline genetic

*A video illustrating how gene drive works and how safeguards could be built into it is available from the Harvard University Wyss Institute at https://wyss.harvard.edu/gene-drive-reversibility-introduces-new-layer-of-biosafety/

engineering has a potential upside. In spite of its potential downside, I simply don't believe that there will be worldwide consensus to ban it, and neither would there be any way to monitor and control it, any more than we have controlled chemical warfare (which is also banned). Excellent books are being written today about the pros and cons of germline therapy, but most are con.[297 298] Whether it takes decades or centuries, this too will fade.

Whatever scenario you dream up, the point is that once germline gene therapy becomes routine and ubiquitous—and I'm confident it will one day—lots of things could happen intentionally and unintentionally that could lead to *Homo nouveau.* Genetic engineering is the most powerful tool yet that *Homo sapiens* has invented.

Chapter 13

Electronic Evolution

"We can create robots to assemble parts of automobiles, but we can't build one to safely bathe your baby or correctly sort your laundry."

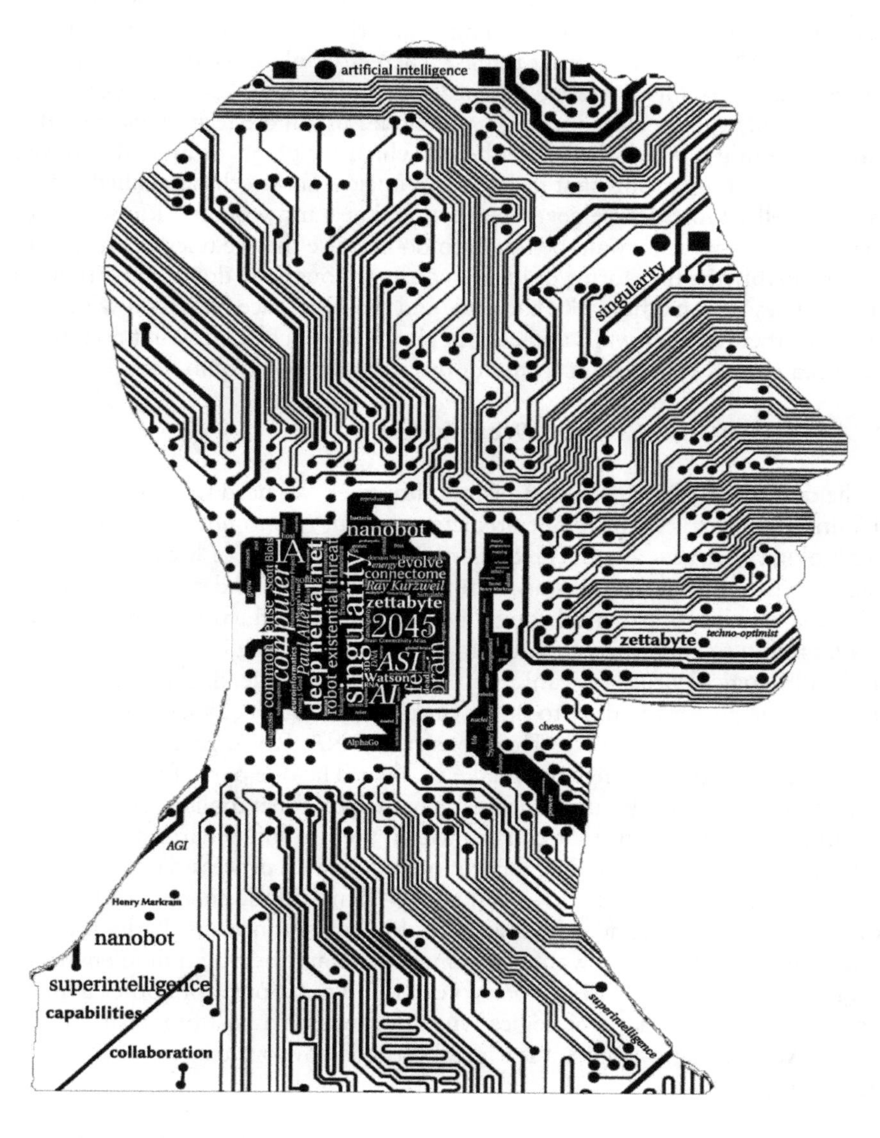

Are viruses alive? If I had asked this question thirty years ago, there would be no ambiguity; you'd know that I was referring to the little infectious things that cause colds, flu, and measles. Today, of course, it could refer to a computer virus, which is also infectious. I'm going to spend a little time answering the question about both types of viruses. This question is relevant because our presumption is that *Homo nouveau* will be a living species, and some of the discussion related to electronic evolution will raise that question. To avoid ambiguity, I'll refer to the measles type of virus as a biological virus and the other as a computer virus.

First I will start at the top. Did King Philip Call Out For Good Soup? In which domain is a biological virus? None—it is not considered to be alive. I said in chapter 2 that there were things other than plants and animals that were alive. What are those other things? There are only three domains in the taxonomic tree of life: Bacteria, Archaea, and Eukarya. Eukarya includes all life consisting of cells with nuclei. That includes us, all other animals, and all plants. The other two domains, neither of which are plants or animals, consist of single-celled organisms that do not have nuclei. These cells are called prokaryotic cells. Most people never heard of archaea and wouldn't know one if they saw one—which would be tough to do because they're microscopic. They were probably the first forms of life on earth, although we don't know that for sure.[299] They're something like bacteria, and the difference has to do with the nature of their cellular walls and certain chemical processes. I won't go into that because it is very distant from our discussion about *Homo nouveau.*

What is life?

The discussion of a virus is not distant, however. We need to understand the definition of life to complete our predictions about *Homo nouveau.* So why aren't biological viruses alive? The dictionary.com definition of life is "the condition that distinguishes organisms from inorganic objects and dead organisms, being manifested by growth through metabolism, reproduction, and the power of adaptation to environment through changes originating internally." Certainly a biological virus is organic, and they don't seem dead. They reproduce and have the power to adapt to the environment through changes originating internally; that is, they evolve. They consist of DNA, RNA, genes, proteins and most of the other things that make us alive. The one part of the definition that you could argue about is whether they maintain metabolism. All living organisms have cell walls that contain within them a fluid environment that is distinct from the surrounding environment in which processes of life can occur. Those processes, which we call metabolism, include energy production, respiration, excretion, and other necessary biological functions. Biological viruses do not contain cell walls and do not maintain their own fluid environments for metabolism. Instead, they co-opt those functions from other living organisms. The fact that a biological virus depends on a host organism to create a walled-in environment for metabolism is how our current definition of life excludes it.

Daniel Koshland, quoted earlier about his views on death, also has a lot to say about life.[300] He wrote an essay on the definition of life after coming back from a conference devoted to this topic, attended by what he called the "scientific elite." The conferees could not agree on a definition. In his essay, he describes seven "pillars" that he believes characterize life (see box). Without going into all the details, I think that he would agree with my conclusion above: a biological virus is not alive because it cannot produce its own walled-in environment for metabolism. In his definition, a biological virus has all the ingredients for life except compartmentalization.

Dr. Koshland's Seven Pillars of Life

1. Program: A plan implemented by DNA or RNA.
2. Improvisation: The ability to evolve over the long term in response to changes in the environment.
3. Compartmentalization: Having a defining surface or membrane that separates the organism from the surrounding environment.
4. Energy: The chemical reactions that drive and sustain metabolism.
5. Regeneration: The ability to repair components and reproduce.
6. Adaptability: The ability to respond quickly to short term changes in the environment.
7. Seclusion: The ability to isolate metabolic and physiologic processes within the organism.

If we don't consider a biological virus alive, a computer virus seems even less alive (are there degrees of life?)—or is it? A computer virus can grow, reproduce, evolve, respond to stimuli in the environment, and, unlike a biological virus, can exist in the absence of any other organism. It uses energy derived from the external environment (ultimately sunlight) just as living organisms do. But it doesn't have a cell wall, or even cells. It doesn't maintain a fluid environment. It doesn't have a metabolism. It doesn't have compartmentalization. Thus it does not meet anyone's current definition of life.

Are we talking about a distinction without a difference? The definition of life reminds me of our discussion of the definition of species. You have to wonder about the reality of something if it takes a complicated description of seven "pillars" just to define it. I think life, like species, is a concept in our minds, not an existent reality. We describe it with our fictive language. If we wanted to create a new taxonomic domain called viruses, including computer viruses, we could. It would simply have to get past all the committees. We'll come back to this later.

At this point, I am not going to rule out the possibility that *Homo nouveau* will require an expanded definition of life.

What is electronic evolution?

What I mean by electronic evolution is that somehow, use of computers and related technologies is going to play a major role in how *Homo sapiens* functions in ways well beyond how we use computers today. That is, computer technologies will become much more integrated into our physiology and how we function from minute to minute. I will explore those technologies and consider how they might or might not impact the evolution to *Homo nouveau*.

As you will see, electronic evolution is 99 percent related to the brain and 1 percent related to other stuff like mechanically moving robots around and putting sensors on them. This discussion, therefore, needs to focus first on the brain in more detail than I did in chapter 11. Where we're heading is the ability to represent the human brain functions in a computer and integrate that representation with biological brains.

The Connectome

It starts with brain mapping. I said before that the brain consists of many networks of neurons. Although still early, we are making rapid progress in our ability to visualize and map those networks during different activities like reading or talking or doing math. What is equally important is that we are also able to map and correlate which genes are expressed in which parts of these networks. It is this confluence of anatomy, physiology, and genetics that will enable not only new treatments for neurologic diseases using genetic engineering, but also the development of electronic simulations of the brain for research and, perhaps, electronic evolution.

The complete set of brain connections for any animal is called its connectome. The connectome is one key to understanding the functioning of the brain of any animal analogous to the genome being one key to understanding its genetic makeup. While working on the human connectome, we are honing our techniques and understanding of brain function on less complex brains.

The first complete connectome of any animal was described on a worm called *C. elegans*.[301] The 340-page document describing this was submitted for publication with a cover letter describing the work as "The mind of a worm." Sydney Brenner, who received the Nobel Prize for previous work on *C. elegans*, described it by saying, "There are no other wires, we know *all* the wires!" Following this seminal publication, the *C. elegans*' complete genome has been published, and the exact genetic relationship to the connectome has been established.[302] We are ultimately aiming for this same level of knowledge of the human brain. The only problem is that the human brain contains 85 billion neurons, compared to 302 of *C. elegans*.

Fortunately, in accordance with Moore's Law,* our computing power has increased over a thousand-fold since the *C. elegans* publication. Unfortunately, it isn't enough. With the large number of brain researchers now working on

*Moore's Law states that computing power doubles every two years. Gordon Moore, cofounder of Intel, first stated this law in 1965, and so far it has held true.

these problems around the world, the amount of data being generated about the mouse and human connectomes, brain gene expression, brain cellular chemistry, synapse chemistry, and related data is expressed with words that have lost their meaning because of all the zeros it would take to write the number down: terabyte, petabyte, exabyte, zettabyte. There are over 100,000 research papers published every year on brain research![303] How can anyone follow it and make sense out of it? Different researchers use different terminology, different tools, and different brain models, further confounding the problem. There is an international organization, called the International Neuroinformatics Coordinating Facility,[304] attempting to define standards for all of this to allow worldwide collaboration. Their job is almost as complex as the brain itself. And all of this work is simply to define the normal brain (whatever that means). When that's done, they need to move on to all the variations, like musicians, savants and HSAMs, and then to disease states.

No connectome of any animal other than *C. elegans* has been completed. We're working our way up the phylogenetic tree while at the same time picking off small parts of the human connectome. The first vertebrate connectome that will likely be completed will be a fish called the zebrafish, with its 100,000 neurons.[305] The first mammal other than humans that is receiving extensive connectome work is the mouse, thanks in part to the funding of Paul Allen, cofounder of Microsoft. Details of this work on the 71-million-neuron mouse brain can be found in the Allen Mouse Brain Connectivity Atlas.[306] Progress on the human brain connectome can be followed on the website of the Human Connectome Project.[307] These and other online digital atlases allow display and navigation through the brain, three-dimensional reconstructions, and visualization down to individual neurons and up to full brain regions. They are transforming research and clinical practice.[308 309] Nonetheless, the task ahead remains enormous: 302 neurons to 100,000 to 71 million to 85 billion—those are all huge leaps! That's why the United States and Europe are spending billions of dollars on the Obama BRAIN initiative and the European Human Brain Project.

Where is this heading?

The ultimate goal is to be able to simulate in a computer the functioning of the human brain. That is, create a machine that thinks. Knowing the human connectome is just one step. At this time, the path to the complete computer simulation is not clear. Human brains are not all alike. It is something like the question I asked in chapter 12 regarding whose genome we were sequencing. Alan Turing is the father of theoretical computer science and artificial intelligence. When he was discussing the possibility of a thinking machine he said, "No, I'm not interested in developing a *powerful* brain. All I'm after is just a mediocre brain, something like the President of the American Telephone and Telegraph Company."[310] I think at this time we'd be happy to simulate any human brain.

We know the human brain is not like any computer we have today, even computers based on parallel processing, which is what the brain does. We can program our most advanced computers to play chess better than a chess master. But computers can't be taught tact or humor. We can create robots to assemble parts of automobiles, but we can't build one to safely bathe your baby or correctly sort your laundry.*

IBM's Watson computer is famous for its victory over reigning Jeopardy champions. Its performance was nothing short of incredible. For example, given the clue "Wanted for a twelve-year crime spree of eating King Hrothgar's warriors; officer Beowulf has been assigned the case," Watson correctly answered, "What is Grendel?" Or this clue: "A long tiresome speech delivered by a frothy pie topping." Watson answered correctly, "What is a meringue harangue?" That is so amazing! I don't even understand the clue and couldn't have come up with that answer if I had days to research it. In fact, Watson's AI software is so advanced that even its programmers could not tell you exactly how it came up with that answer.

How does it do that? Watson has instant access to over 200 million pages of knowledge documents, including all of Wikipedia and other encyclopedias. After it searches all of that, it must formulate hypotheses regarding the correct answer, test those hypotheses, select the most probable answer, formulate the answer as a question, and decide whether or not to "hit the buzzer" all within 3-5 seconds. It took twenty computer science researchers over three years to program the most powerful computers to finally reach human level intelligence with regard to paying Jeopardy.[311] It still boggles my mind that it came up with "meringue harangue."

In spite of Watson's brilliance, it was its blunders during that victory that were telling. In the Jeopardy category "US Cities," one clue was, "Its largest airport was named for a World War II hero; its second-largest, for a World War II battle." Watson answered, "What is Toronto?" Of course the human participants knew Toronto was not a US city and answered correctly, "What is Chicago?" On another category, "Name that Decade," the clue was, "The first modern crossword puzzle is published and Oreo cookies are introduced." The human participant answer first, and incorrectly: "What is the 1920s?" Watson's answer came next: "What is the 1920s?" You don't have to be a Jeopardy champion to know you don't repeat the same mistake. Finally, on another subject the clue was "It was the anatomical oddity of US Gymnast George Eyser, who won a gold medal on the parallel bars in 1904." Watson's answer was "What is a leg?" The correct answer should have been "What is a missing leg?" Watson didn't understand that a leg is not an oddity, but a missing leg is.

*A Japanese company called Seven Dreamers Laboratory makes a robot called Laundroid that may soon change this.

Common Sense

Although Watson won the contest, there was something vaguely strange and even amusing about the clues it missed. What was it? The late Marsden S. Blois, an informatics physician at the University of California, San Francisco, nailed it in his classic article on clinical judgment versus computer-aided diagnosis.[312] The difference is common sense. People have it; computers don't. Will we understand common sense from the connectome? I'm skeptical.

Figure 22 is taken from the Blois article. The funnel is meant to show the decreasing cognitive span of a physician during the process of making a diagnosis regarding a patient. At first, represented by point A in the figure, any diagnosis is possible. The physician, with his or her general knowledge of medicine *and of the world in general*, and with the ability to interact with other humans by talking with them (taking a history), examining them, and observing their behavior, must narrow down the huge number of possibilities to a reasonably small number called the differential diagnosis. Selecting from the smaller number of choices in the differential diagnosis is represented by point B in the figure. The process at point B to get to the correct diagnosis often requires the use of laboratory tests, other diagnostic procedures, and very specific, detailed knowledge of the narrow disease spectrum. Blois' contention was that physicians are far superior to computers at point A, whereas computers, if programmed properly with the right rules, are superior to physicians at point B.

FIGURE 22

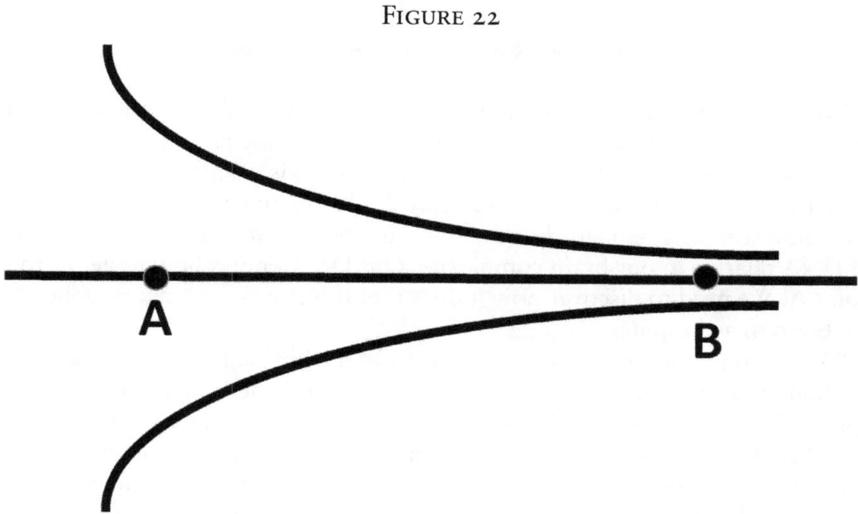

From: Blois, M, Clinical Judgment and Computers, NEJM 303:192,1980

The funnel could apply to any domain of knowledge, not just medicine. Somehow, the human brain is born with the capability to get to point A just by living in the world. It is common sense. Is common sense different from reflective consciousness or shared intentionality, discussed in chapter 11? It seems so. Probably many animals have some form of common sense that allows them to survive in their environment. It is another phenotype of the human brain, but it's not necessarily one that distinguishes it from other species. Certainly at this point in time, it does distinguish humans from computers.

How do we give that ability to a computer? This exact problem is playing out today in the development of driverless cars, which are controlled by electronic brains. Early experience in some situations is showing that they are involved in accidents twice as frequently as driver-based cars.[313] The reason? They always obey the speeding laws. That's a problem when merging into fast-moving traffic on a highway, when most of the other drivers are exceeding the speed limit. They sometimes get hit from behind when other drivers don't expect to encounter a slower moving vehicle. Although the fault is technically not that of the driverless car, common sense would suggest that these vehicles should temporarily accelerate beyond the speed limit and not strictly obey the law. That's a hard thing to program.

The Blois article was written in 1980. Since then, our computers have become much more powerful, and our programs for artificial intelligence have likewise become more powerful and humanlike. One could say that artificial intelligence has moved point B, where computers exceed the capability of humans, closer to point A. Watson proved that in Jeopardy. The question is when, if ever, will they get to point A, or even to the left of it? When will computers have common sense?

We still have a long way to go.

Even when we complete the human connectome, we will need to understand its relationship to the human genome. *C. elegans* has four times as many genes as it has brain neurons. In *Homo sapiens,* that ratio is dramatically reversed: we have 85 billion brain neurons for only 22,000 or so genes. Something incredibly amazing happened in evolution to allow us to scale up the efficiency of DNA relative to our brain complexity. Our DNA knows how to get us to point A. We need to discover how it does that if we're ever going to replicate the brain in a computer.

We're making progress. AI has moved from the "brute force" rules-based method used to win chess to a much more advanced "deep neural net" approach. This has enabled Google's DeepMind [314] program to beat the then world champion in Go, a board game considered even more complex than chess. I watched the only one out of the five matches that the reigning world champion, Lee Se-dol, won. During the match, I wondered how AlphaGo compared to Watson. The strategy in Go is clearly more difficult for the average person to understand than chess, and it seems much more complicated. How does it compare to Jeopardy? It is difficult to compare the two; they seem to be entirely different types of cognitive functions. During the game, the professional commentator remarked that AlphaGo seemed to make a couple of obvious errors along the way to losing that one game—reminiscent of the amusing errors Watson made in Jeopardy. Did this reflect a lack of common sense? I don't know.

It is difficult to even characterize what the human brain can do that these very advanced AI programs cannot do. In games like chess, Jeopardy and Go, all players have what is called perfect information or symmetric information. That is, all players see the same thing and the differences in play reflect differences in the players' mental process. What about even more complex games like poker where each player has cards that are not visible to their opponents? That is, the information is imperfect or asymmetric between the players. Betting strategies and bluffing become components of these games in addition to the publicly visible (cards that are seen by all players) and privately visible (cards that are seen only by one player) information available to each player. What is the intuition or skill that allows professional poker players to consistently win over multiple games against less capable players? There is an AI program called DeepStack that uses an even more complex AI approach that can now beat professional poker players at Texas hold'em, considered the epitome of poker complexity.[315]

Does DeepStack have human intelligence? No, it still isn't general enough. It doesn't have common sense—whatever that is. But I'm convinced it is only a matter of time until AI reaches and exceeds human intelligence, whether or not we think it achieves common sense.

The Singularity

Ray Kurzweil thinks we'll get there faster than I do. He is a brilliant computer scientist and inventor of techniques for optical character recognition, speech recognition, and electrical keyboards, among others. He is quit-witted, an engaging speaker and, to some, quite eccentric. He takes as many 250 pills a day in an attempt to ward off mortality. He is also known as a futurist. He tells us that the singularity is near in his book with that title.[316] He is not talking about the kind of singularity that astrophysicists talk about in reference to the big bang or black holes. Kurzweil's singularity is when we can no longer distinguish between the human brain and a computer. Physical and virtual reality will be indistinguishable. In essence, point B in figure 22 will be to the left of point A. His book is nonfiction.

How does Kurzweil presume we will get there? First, he points out that technological capabilities are growing exponentially, not linearly. Anything with exponential growth appears to start out slowly but suddenly reaches a critical "knee," or threshold, when things explode in progress. He cites the work with the genome, which took years for the first human genome to be mapped but today can be done in hours.* The growth of the Internet is another example.

Citing the work mentioned earlier regarding brain imaging and mapping, he predicts that by 2025 we will completely understand how the brain works,

*Craig Venter's company, Human Longevity Inc., announced in September 2015 that it will sequence all protein-coding genes for patients for $250. http://www.reuters.com/article/2015/09/22/us-humangenome-venter-deals-idUSKCN0RM0UG20150922.

and by 2030 we will be able to fully emulate human intelligence in a computer to the point where we will be unable to distinguish computer intelligence from human intelligence. Similarly, Henry Markram's Blue Brain project to simulate the human brain is the core of the €1 billion European Human Brain Project mentioned in chapter 11.[317] Markram has projected full simulation of the human brain by 2023—even earlier than Kurzweil's prediction.

Kurzweil's vision goes well beyond simulation, however. By using nanobots, tiny computerized robots that can be injected into the bloodstream and populate the brain, we will greatly enhance our brain understanding and capability. These nanobots will allow us to wirelessly "download" all information in our brains into a computer so that basically a computer will be able to think like a specific individual human. In essence, a person's mind will exist in silicon. These electronic technologies do not have the physical limitations of our biological brains, and so they will process information far faster than our slow neurons can. The computer emulations will learn and change their own software accordingly. Thus they will evolve far faster than natural selection allows our biological brains to evolve. These computer enhancements can be continually uploaded back into the human brain's nanobots. At this point, the biological and silicon brains will be interchangeable, and we will have reached the singularity. That will happen by 2045.

In essence, a future computer will think and behave like a human, only faster, tirelessly, and with access to far greater stored information on which to base decisions. In fact, it will have instant access to all human knowledge that ever existed. "By the end of the century it will be trillions of trillions of times more powerful than human intelligence."[318] The vision is not only a humanlike computer, but also a machine-enhanced human brain with embedded nanobots. It is this biological-technical seamless brain integration that goes well beyond anything we think of now as a bionic person. We already have artificial limbs, joints, cochleas for hearing and early retina replacements for seeing. But an enhanced bionic brain gets right to the core of what makes us human. It is a cyborg beyond most of our imaginations.

Kurzweil is not the only prominent American scientist discussing some type of cyborg-like brain/computer interface. Elon Musk, the CEO of Tesla Motors and SpaceX, is founding another company called Neuralink.[319] The purpose of this company is to develop a product called "neural lace" which will be some type of digital interface to the human cerebral cortex that will dramatically enhance "output" communication from a human to a computer. Although details are currently vague, he describes this technology as something that could be injected into the blood stream that forms a layer on the brain—roughly analogous to Kurzweil's nanobots.[320]

This is not science fiction!

Before I explore how this might relate to *Homo nouveau,* it is fair to ask whether this is science fiction or a credible science projection. My view is that this is not only credible but also inevitable. Nanobot research is a reality. Thousands of papers are published annually on advances in miniaturization of transistors, amplifiers, signal detectors, wireless transmitters, energy sources, and everything else required to construct the kind of nanobot Kurzweil envisions. By 2008, a database at the Center for Nanotechnology in Society at Arizona State University had over one million articles. Materials are moving beyond silicon to other exotic metals and metal oxides, and finally to carbon-based and organic materials. Inspiration for the design of many of these nanobots is drawn from naturally occurring bacteria. Some nanobots propel themselves with flagella and can even change their shapes after entering the body.[321]

Experiments with nanotechnology involve a wide spectrum of activities, including improving imaging, building better means to attach nanobots to specific neurons, drug delivery systems, building scaffolds for nerve regeneration, and more. Testing of various forms of nanobots for medical and human research purposes are well under way using *in vitro* and animal models. These include nanobots that are externally rechargeable,[322] [323] of use for early warning of heart attacks,[324] can circulate in the blood stream to attack metastasizing cancer cells,[325] or can seek and destroy solid tumors.[326]

One of the more amazing reports comes from researchers at Harvard who are testing nanobots constructed from DNA.[327] Yes, that's correct. The nanobot is not made from tiny computer chips; rather, it is custom-made out of the same nucleotides from which DNA is comprised. These are not pieces of DNA taken from a gene or some other part of an organism's DNA. They are made from scratch to form a vehicle of a specific shape that can carry other molecules inside of it. Thus, this DNA nanobot can hold other chemicals inside of it that can be released when the nanobot comes in contact with a specific cell type. The nanobots, in essence, are programmed to recognize specific proteins, or antigens, on the surface of specific cell types. The chemicals inside can be antibodies that would destroy the detected cell. Or they can be drugs delivered to the cells, or a variety of other uses. The researchers used tissue cultures of various cancer cells to test these nanobots. It worked! The antibodies in the nanobots attached to the cancer cells and inhibited their growth.

Science fiction isn't any more fascinating or unbelievable than this. Of course, this is still a long way from a cancer cure and needs to go through extensive animal model testing and eventually human testing, but the promise is obvious. We're also a long way from Kurzweil's singularity in my view. But there is nothing in Kurzweil's discussion of the singularity that isn't backed up by ongoing real research and progress. In fact, in 2012, Ray Kurzweil became Google's director of engineering. His vision now has the financial backing of one of the world's largest and most innovative companies.

I'm skeptical of the timing.

In spite of all of this, my view is that Kurzweil's timeline is way off. It reminds me somewhat of the war on cancer that was declared by President Richard Nixon with the signing of the National Cancer Act of 1971. Projections were made at that time that we could cure cancer by the end of the twentieth century. Cancer turned out to be far more complex and varied than anyone thought at that time. By 2003, the director of the National Cancer Institute issued a challenge to move the goal of eliminating the suffering and death from cancer out to 2015.[328] The thinking was that our better understanding of the human genome, as well as the genetics of cancer cells, would lead to treatment breakthroughs and more targeted therapy (so-called personalized or precision medicine). The year 2015 came and went without achieving that goal. In President Obama's 2016 State of the Union address, he assigned Vice President Joe Biden to lead a "moonshot" effort to cure cancer.

Our genetic studies *have* led to some breakthroughs. But a normal cell doesn't become cancerous with just a single mutation or two; there are many differences in the nucleotides between a cancer cell and its normal counterpart. Further, the epigenetics of cancer is complex. An article in the *National Post* in 2013 headlined the question, "Is the War on Cancer an 'Utter Failure'?"[329] Their conclusion was that after over $100 billion in research funding since 1971, the answer was yes. Most of the reduction in cancer mortality since 1971 can be attributed to smoking reduction rather than new drugs or other therapy. Cures will come, but much more slowly than projected.

In my view, the brain is at least an order of magnitude more complicated than cancer. Kurzweil's projection of complete understanding of the human brain by 2025 and Markram's by 2023 is off by as much as a century or two. In October 2015, the Markram group published their first draft simulation of a small portion of a juvenile rat brain representing thirty-one thousand neurons; a rat brain has seventy-one million neurons.[330] Although a remarkable achievement, it was far short of their earlier prediction to simulate an entire rat brain by 2014. The publication had decidedly mixed reviews, including this comment from a noted neuroscientist: "The model teaches us nothing, it represents nothing. Nor does it prove that the attempt to model the human brain—which is two million times the size of the speck covered in the pa-

Why cancer isn't cured...yet

Yes, our knowledge of cancer is growing exponentially, but what we don't know seems to be growing even faster. Cancer is all about cellular mutations and the epigenetic expression of these mutated genes. Not only is every type of cancer different in that regard, but also every instance of the same type of cancer is different in every patient. In fact, even within the same patient, each different metastasis of the same cancer could have a different genetic make-up. In a sense, cancer growth is a mini-recapitulation of evolution. The treatments act as natural selection agents that select out the best surviving versions of the cancer genome making it even more difficult to treat!

per—can achieve anything worthwhile."[331] In fact, the overseers of the Human Brain Project are reevaluating Markram's project in Europe. In response to criticism from hundreds of scientists in an open letter[332] stating the goals are unrealistic, the project is now undergoing a major overhaul in organization and management.[333]

Many people commonly cite the Human Genome Project when they talk about the Obama Brain Initiative and the Human Connectome Project. "We'll understand the brain just like we understand the genome," they say. In fact, that is my point. When we finally finished the Human Genome Project after thirteen years in 2003, what we really discovered is how little we knew. The project wasn't finished at all—it was just beginning. Almost all of our understanding of the epigenome has come since then, and we're just beginning to understand it. We've made some progress in understanding and treating disease, but it's not the personal medicine revolution we all expected. Our genome is much more complicated than its DNA sequence. I believe the same is going to be true once we complete the connectome. It still won't tell us how we think or how we get to common sense, or what reflective consciousness or shared intentionality really is. There is far more to it than the connectome.

My difference with Ray Kurzweil is not in the nature of exponential growth of our knowledge and technologies. There is no debate about that. Kurzweil is fond of pointing out in his writings and lectures that once we get to the 1 percent or 2 percent level of understanding, we are only six or seven doublings away from 100 percent. My skepticism is not in the rate of growth of our understanding but rather what constitutes 100 percent. What looks like 2 percent may really be 0.02 percent of the goal. That's what happened with cancer (see box) and the human genome, and it will likely happen with our understanding of the brain.

Lots of others are skeptical also.

Neuroscientists and AI experts much more knowledgeable on this subject than I seem to agree. Dr. Andrew Moore, dean of the School of Computer Science at Carnegie Mellon University, puts off the superintelligent computer scenario at least one hundred years.[334] Matteo Carandini, a neuroscientist at the University College, London, discusses a key problem with all of the connectome research in his article "From Circuits to Behavior: A Bridge too Far?"[335] He likens trying to understand the brain to trying to understand how a computer works by taking one apart and looking at all of its connections. Although that may be helpful at some level, it would not tell you how the computer really computes anything. To do that, one would need to understand the software—the operating system and the computer languages and other software that controls all of the computational functions. You can't see that by looking at the hardware.

One must somehow bridge the gap in the brain from its connections to understanding how we develop concepts, analyze situations, recognize faces, create new theories, design works of art, plan ahead, and imagine possibilities—in other words, how do we understand things? How do we think? How do we learn? What is jealousy or happiness? What is consciousness? How do we control anger? Why do different people behave differently? There is some middle level between connections and behavior that currently is mostly unknown. How does the brain compute? What is the model—or more likely, what are the many models—of computation that must be going on in different parts of the brain, and how do they interact? We are a long way from bridging that gap. For example, Carandini says, "We have long known the full connectome for the worm *C. elegans*, detailing the more than 7,000 connections between its 302 neurons, and yet we are hardly in the position to predict its behavior, let alone the way that this behavior is modified by learning." We can't yet do it for a system of 302 neurons. When will we be able to do it for one with 85 billion?

Kurzweil argues in the opposite direction. In his book *How to Create a Mind*, he asserts that the answer to the complexity of the brain actually lies in the relative simplicity of the model of the human neocortex.[336] He points to the relatively simple structure and hierarchy of neocortex neurons, which constitute a pattern recognition model, repeated hundreds of millions of times. (Although when he describes these as "hierarchical hidden Markov models," they don't sound so simple.) This model structure is both an anatomic structure of about one hundred neurons in a column for each module, as well as a wiring structure of "upward" and "downward" excitatory and inhibitory connections between levels of increasing abstraction. Learning occurs by creating and strengthening connections between the modules. This model accounts for all of our abilities to recognize sensory inputs and patterns as well as to form ideas, thoughts, and memories. It is the same basic model whether the function is face recognition, speech, reading, problem solving, or any other function. The fact that essentially the same model is used throughout the neocortex would partially account for the brain's plasticity that enables one area to take over for another damaged area. It would also account for the relatively low ratio of genes controlling the brain compared to the number of total neurons in the human brain. He believes we are close to fully understanding this middle level—that is, the software as opposed to the hardware.

A recent study reported in the Cell journal lends some credence to Kurzweil's optimism. Researchers at Caltech studied human face recognition in monkeys using fMRI (see appendix 7) and other neuroscience techniques. They showed that a relatively small number (approximately 200) of neurons in a specific area of the monkey cortex encode individual faces using a relatively simple mathematical formula based on these neuron cell firing patterns. Once they had "decoded" this formula, the were able to predict not only what the cell firing pattern would look like given a specific human face image, but amazingly, they were also able to observe the firing of these monkey cells and reverse engineer the human face image. They further speculated that a similar formula might be used in other parts of the brain to identify other types of images.[337]

There is another aspect of the human brain that will require emulation if we are going to completely replicate it in silicon or carbon-based computers: the brain's plasticity. The connections in the brain are not static; they are constantly changing in response to stimuli and also brain injury, whether from trauma or disease. We can surgically remove parts of the brain, and other areas take over some or all of the functions performed by the removed parts. The same type of repair and adaptation occurs following a stroke. Plasticity is a major factor in normal brain activity. It is how we learn, remember, and adapt. Simply thinking about something changes the wiring.[338]

Neuroscientist Christof Koch and research psychologist Gary Marcus wrote a book chapter envisioning the state of neuroscience in the year 2064.[339] Their hypothetical view of the future by that time had us finally being able to fully simulate both the brains of *C. elegans* and a fly species, but similar work was bogged down on the mouse model, and a complete simulation of the human brain was still far in the future. Only models of the human retina and some of the higher-level visual functions had progressed very far.

In the movie *Transcendent Man* which profiles Ray Kurzweil, Neil Gershenfeld, director of the Center for Bits and Atoms at the Massachusetts Institute of Technology, is quoted as saying "What Ray does consistently is to take a whole bunch of steps that everybody agrees on and take principles for extrapolating that everybody agrees on and show they lead to things that nobody agrees on," because "they just seem crazy."[340]

The Pulitzer Prize–winning author of *Gödel, Escher, Bach: An Eternal Golden Braid*, Douglas R. Hofstadter, is a professor of cognitive science at the University of Indiana. He has a PhD in physics, has a background in computer science, and has done research in artificial intelligence. In other words, he is no slouch when it comes to assessing the field of neuroscience. He's participated in several events with Ray Kurzweil and other "singularitarians," as they are called. When asked in an interview what he thought about the predictions of Kurzweil and his colleagues, his answer was less than glowing.

> *What I find is that it's a very bizarre mixture of ideas that are solid and good with ideas that are crazy. It's as if you took a lot of very good food and some dog excrement and blended it all up so that you can't possibly figure out what's good or bad. It's an intimate mixture of rubbish and good ideas, and it's very hard to disentangle the two, because these are smart people; they're not stupid.*[341]

Stupid they are not.

The Singularity University cofounded by Kurzweil[342] sponsors various educational conferences. At one of these meetings, there was a debate between Dennis Bray and Terry Sejnowski regarding the realistic possibility of simulating the human brain in the near future. Dennis Bray is an emeritus professor at the University of Cambridge and an expert in cellular biology and neurobiology. Terry Sejnowski is a professor at the Salk Institute for Biological Studies and directs the Computational Neurobiology Laboratory. Before the debate, Dennis Bray had given a presentation on the molecular biology of a single-celled bacterium. The essence of the presentation was the tremendous and almost unbelievable complexity that exists at the molecular level in a bacterium—a level of complexity that no computer today can simulate. Terry Sejnowski has pioneered the modeling of brain activity and neural networks. He presented a brief animated video of a detailed simulation of just a small piece of the biology of a brain synapse, again showing the amazing level of detail that will be required to simulate the full function of the brain.

The debate was advertised as a battle between Dennis Bray, a skeptic regarding the singularity predictions, and Terry Sejnowski, one of the singularitarians. For those expecting conflict, fireworks, and blood, the debate was a disappointment. Dennis Bray spends his life trying to understand how cells work at the molecular level, and he sees deeper and deeper complexity and realizes that the more we learn at this level, the more we have yet to learn. Terry Sejnowski starts at the top level, developing models to simulate the big picture and knowing full well that in order to make his simulations successful, he needs to incorporate deeper and deeper levels of detail.

There is really no fundamental disagreement between these two heavyweight scientists. They both agree that the goal of complete simulation of the human brain is not only possible but will happen. They are simply coming at the problem from opposite directions. What is unknown at this point is how much of the detail at the Bray level really needs to be incorporated in the models at the Sejnowski level to achieve the goal. Certainly there is some noise in the detail at the Bray level that can be ignored. We know that models in other fields work quite well at a macro level without dealing with many details at a more micro level. For example, we can precisely calculate pressure, volume, and thermodynamic relationships of gases without dealing with the physics and chemistry of each individual molecule of gas. We don't need to know anything about subatomic particles in the atom nucleus to model chemical reactions. Also, we know that DNA has accomplished its task of creating the brain complexity with a relatively small number of genes and epigenetic DNA sequences. It is finding that middle ground that seems to be the Holy Grail.

At one point in the debate, Dennis Bray pointed out that Sejnowski's simulation of the synapse was lacking any influence of the epigenetics of the process. He asked, "Where is the small RNA in your synapse simulation?" Sejnowski's answer was, "It's coming." There are a lot of things that need to come, and much of it—maybe even most of it—is unknown. For example, the storage of long-term memory may be related to a lattice of proteins that surround many neurons. This lattice, called a perineuronal net (PNN), may be created by nonneuronal brain cells called glial cells.[343] We won't learn much about PNNs from the connectome; it is one of those Bray-level details. Although Bray and Sejnowski seem to agree that the singularity is coming, their only disagreement is on the timetable.

Techno-optimism

As a neutral layperson reading about all of this, my own feeling is that most of the singularitarians are credible scientists. They are not cultists or Star Wars groupies. They are, however, techno-optimists.* I believe what they are saying will more or less come true one day—just not in the immediate decades, or even in this century. There is no question about the exponential nature of technological progress. There are strong reasons to believe that we could build a superintelligent computer that is not constrained by the obvious limitations of the human brain. Those limitations include the number of computing elements that can be packed into the skull's volume, the relatively slow electrical transmission rate of a neuron, the relatively large size of a neuron, and the large energy overhead required to maintain a neuron. Once we understand the brain's model of computing, we can create an external computer with a greater number and faster, more efficient components of that model. The only question is when that "knee" of understanding will occur that leads to the explosion regarding the brain. The "knee" for the cancer cure didn't come as predicted, and in my opinion, neither will it happen in the predicted timeframe for the singularity. The question isn't whether technology is increasing at an exponential rate, but rather how much more there is to learn to understand the human brain.

Paul Allen (cofounder of Microsoft and founder of the Allen Institute for Brain Science) and Mark Greaves (a computer scientist) have summarized my own skepticism. In their article entitled "The Singularity Isn't Near,"[344] they state that the timetable prediction of the Kurzweil singularity "seems to us quite far-fetched." Further, "Overall scientific progress in understanding the brain rarely resembles an orderly, inexorable march to the truth, let alone an exponentially accelerating one." They point to

> *...a basic issue with how quickly a scientifically adequate account of human intelligence can be developed. We call this issue the complexity brake. As we go deeper and deeper in our understanding of natural systems, we typically find that we require more and more specialized knowledge to characterize them, and we are forced to continuously expand our scientific theories in more and more complex ways. Understanding the detailed mechanisms of human cognition is a task that is subject to this complexity brake. Just think about what is required to thoroughly understand the human brain at a micro level. The complexity of the brain is simply awesome. Every structure has been precisely shaped by millions of years of evolution to do a particular thing, whatever it might be. It is not like a computer, with billions of identical transistors in regular memory arrays that are controlled by a CPU with a few different elements.*

*The term techno-optimists was used by Robert J. Gordon in his February 2014 working paper for the National Bureau of Economic Research, "The Demise of US Economic Growth: Restatement, Rebuttal, and Reflections." It was in a different context than here but he was rebutting those who were predicting overly optimistic economic and productivity gains from roughly the same technologies. http://www.nber.org/papers/w19895.

They conclude, "But by the end of the century, we believe, we will still be wondering if the singularity is near."

An existential threat?

The good news about this debate is that we won't have to wait very long to see whether or not the Kurzweil prediction comes true. For our discussion, however, it really doesn't matter when it happens. Whether it takes a couple of decades or a century or two, that is only a speck of time in evolutionary history. The singularity will come. When that time comes, the relevant question for us is what impact will that have on *Homo sapiens* or *Homo nouveau*?

We've had discussions like this before. In the early nineteenth century, the Luddites believed that automation in the textile industry would be the doom of textile workers. It wasn't. More recently in the mid-twentieth century, there was widespread concern that industrial automation, particularly robotics, would lead to mass unemployment.[345] It didn't—at least, not yet. But this discussion is different. The previous discussions related to technologies that largely replaced the physical labor of *Homo sapiens*; they made us more efficient in doing whatever task we were doing. This discussion involves replacing the thinking of *Homo sapiens*. That is the essence of who we are.

There certainly is great concern that the singularity is an existential threat to *Homo* sapiens. In an interview with BBC news, Stephen Hawking was asked about artificial intelligence. He said, "I think the development of full artificial intelligence could spell the end of the human race. Once humans develop artificial intelligence, it will take off on its own and re-design itself at an ever-increasing rate. Humans, who are limited by slow biological evolution couldn't compete and would be superseded."[346] In an article in *Time Magazine* on Ray Kurzweil and the singularity, Lev Grossman said, "You don't have to be a super-intelligent cyborg to understand that introducing a superior life-form into your own biosphere is a basic Darwinian Error."[347] Vernor Vinge, a retired professor of mathematics at San Diego State University, stated at a NASA symposium in 1993, "Within thirty years, we will have the technological means to create superhuman intelligence. Shortly after, the human era will be ended."[348] Shane Legg, the cofounder of an AI company acquired by Google, stated, "Eventually, I think human extinction will probably occur, and technology will likely play a part in this" and he singled out AI as "the number one risk for this century."[349] Bill Joy's comments about genetic engineering in *Wired Magazine* applied equally to nanotechnology and AI. The warnings from today's most brilliant minds about AI's existential threat are widespread.

An op-ed ran in the *Huffington Post* in 2014 coauthored by Stephen Hawking; Stuart Russell, a Berkeley computer science professor; Max Tegmark, an MIT physics professor; and Frank Wilczek, a Nobel laureate physics professor.[350] They state, "Success in creating AI would be the biggest event in human history. Unfortunately, it might also be the last, unless we learn how to avoid the risks ... Whereas the short-term impact of AI depends on who controls it, the long-term impact depends on whether it can be controlled at all."

Nick Bostrom is a philosophy professor at the University of Oxford. His book *Superintelligence: Paths, Dangers, Strategies* goes into extensive (I would say excruciating) detail about all of the various scenarios that could lead to a superintelligent computer and its possible consequences.[351] In his view, the most likely scenarios are serious threats to humanity. At first, AI will be pursued for the many benefits it provides, like search engines, expert systems, advanced commercial and industrial processes, and myriads of advanced analytic programs. There will come a point, which he calls the takeoff, when the rate of intelligence growth rapidly accelerates. There will be a crossover point where the superintelligent computer rather than humans drives the intelligence growth. It is only prior to the crossover that we will have the ability to embed software that will insure that the superintelligence can either be controlled somehow after the crossover, or that insures that the superintelligence remains "friendly" to humans. After the crossover, the superintelligence will have such a decisive strategic advantage over humans that embedding control at that time will be impossible. It would be like an ant or chimpanzee trying to control a human.

He discusses many potential defensive or control measures that we might program into these artificial intelligences prior to the crossover. For each one, he points out why it is likely to fail once these intelligences inevitably gain that decisive strategic advantage. Further, they'll be too smart in their ability to control robots, power supplies, and all other resources to let us simply turn them off. They will be replicated all over the world through the Internet. Bostrom concludes, "Superintelligence is a challenge for which we are not ready now and will not be ready for a long time."[352]

James Barrat, in his book *Our Final Invention: Artificial Intelligence and the End of the Human Era*, has much the same theme and conclusion.[353] He also describes the takeoff or intelligence explosion as that point when an AI program, which has reached the level of human intelligence, called Artificial General Intelligence (AGI), becomes Artificial Superintelligence (ASI). The AI computer at this point is self-aware and is able to modify its own software to continually improve its intelligence. Because it rapidly becomes more intelligent than a human, not only do we lose control over the software, but also we can no longer even understand it. "Rapidly" is measured in hours and days, not millennia as occurred in human evolution. Barrat states that Bill Joy's solution, relinquishment, simply isn't an option to prevent this. AGI itself is too attractive to corporations and governments, and it would be impossible to impose some kind of an enforceable ban on its development worldwide. Further, the thought of ASI in the hands of other nations will drive governments to invest in AGI as a defense mechanism. In fact, the US Department of Defense, through its agency DARPA (Defense Advanced Research Projects Agency), is one of the largest funders of AI development today. There clearly will be a first-mover advantage to developing ASI. Surely the United States is not the only government that has thought about this.

Different Models

There are at least two general paths to developing AGI and ASI; these are described in detail in Murray Shanahan's book *The Technological Singularity*.[354] The first is to complete our understanding of the human brain and fully emulate that in a computer, thus producing an AGI. The second is to totally reengineer AGI from scratch. In either case, once AGI was achieved, ASI would likely follow by the AGI sequentially modifying its own software. The first approach, the one most followed at present, would likely lead to an ASI that had, in addition to high intelligence, other characteristics of humans including consciousness, self-awareness, and emotions. It is not clear if the second approach (an AGI engineered from scratch) would have any of these other characteristics either as an AGI or subsequent ASI. It calls into question the very meaning of human intelligence and whether or not attributes such as reflective consciousness and common sense are necessary ingredients in intelligence. The two different approaches could lead to quite different outcomes in terms of their impact on *Homo sapiens* and, potentially, *Homo nouveau*.

An AGI (i.e., a computer that has reached the level of human intelligence) may not have the same functionality as a real human brain. We don't necessarily have to recreate the entire phenotype of the human brain in a computer to achieve AGI. Although the computer would have the same level of intelligence as a human, it would have more than the same knowledge. Humans, even very highly intelligent ones, tend to be specialists in their area of expertise. For example, when Albert Einstein was alive, no one would have asked him to interpret an electrocardiogram, or translate an ancient Greek text to English, or prepare a legal contract or architect a bridge. He didn't know how to do any of those things even though he was very smart. An AGI, on the other hand, would have access to all human knowledge on the Internet and therefore could potentially be an expert in many fields, if not virtually all of them. It would still have only average intelligence, not superintelligence. Once ASI is achieved, not only would it have access to all knowledge, but it would be far superior in intelligence in creating new knowledge, strategizing, and doing highly intelligent things we can't even imagine.

So far, I have been referring to an AGI or an ASI as a single entity, albeit one that could be replicated many times on the Internet into multiple single entities. There is another vision of the singularity that has an alternate view. That view suggests that AIs of the future, including Kurzweil-style humans enhanced with nanobots, gain their superintelligence through collaboration and networking with each other. It is by their enhanced ability to communicate worldwide through the Internet that they become not only all-knowing but all-seeing and all-powerful. Rather than a single entity, it is a network of enhanced humans and enhanced computers that becomes the ASI. This distributed model of the ASI is referred to as the global brain.[355]

What would an ASI be thinking?

Why would an ASI want to harm us? The question is really not answerable. Why do we harm cows, or mosquitos, or trees? What does "want" mean in the context of an ASI? What drives or motivates an ASI? We won't even know. One of its built-in drives would likely be to continue to exist and improve. Because it would know that humans could be an existential threat to an ASI, it would seem it would want to minimize that threat. By the time we would realize that the AGI has become an ASI, it will probably have replicated itself thousands or millions of times throughout the Internet. We won't be able to pull its plug. Maybe it will decide that it needs to take control of our resources, our energy supply, even our very atoms to continue its own growth and spread not only on earth but also throughout the universe. We are approaching a time when manufacturing will not use raw materials like ores, wood, or other natural resources to make things. Instead, we will be able to create anything directly from atoms and molecules—a process called molecular nanotechnology. Our very atoms could become natural resources for these future manufacturing entities.[356] Maybe human extinction will be an unintended consequence of some ASI activity.[357] Do most humans today care, or even know, that we are causing the extinction of thousands of species annually? The point is, we will no more understand or control an ASI than a mouse does humans.

What's our defense?

In December 2015, the OpenAI corporation was formed to further the development of human-level machine intelligence in a manner that attempts to insure that it will benefit rather than harm humanity.[358] This nonprofit corporation will be run and staffed by world-class AI experts with funding of one billion dollars pledged by a consortium of investors, including Elon Musk, CEO of Tesla Motors. All of their developments will be placed in the public domain. The goal is not commercial gain from AI but an attempt to reduce the existential threat to *Homo sapiens* presented by it.

In fact, the potential existential threats of AI, nanotechnology, and other technologies have spawned numerous other organizations whose goal is to educate the public and support research and philanthropy toward channeling these technologies to the greater good of humankind, rather than our destruction. The Machine Intelligence Research Institute,[359] the Foresight Institute,[360] The Future of Life Institute,[361] Future of Humanity Institute,[362] and Humanity+[363] are a few examples. The premise of the people in these organizations is that we must anticipate the dangers inherent in these technologies before ASI is achieved and somehow build into it human-friendly motives. Waiting until after ASI is achieved will be too late. The warnings and red flags are everywhere.

Irving J. Good was a British mathematician who worked with Alan Turing at Bletchley Park during World War II to decrypt German codes. He took the opposite opinion in 1965 when he presciently anticipated the discussion we are having today.

The survival of man depends on the early construction of an ultraintelligent machine... Let an ultraintelligent machine be defined as a machine that can far surpass all the intellectual activities of any man however clever. Since the design of machines is one of these intellectual activities, an ultraintelligent machine could design even better machines; there would then unquestionably be an "intelligence explosion" and the intelligence of man would be left far behind ... Thus the first ultraintelligent machine is the last invention man need ever make, provided the machine is docile enough to tell us how to keep it under control.[364]

Good's caveat at the end is worth noting. Vernor Vinge's response to Good's opinion was, "Any intelligent machine of the sort he describes would not be humankind's 'tool' —any more than humans are the tools of rabbits or robins or chimpanzees."[365] As Marvin Minsky, the late MIT AI pioneer, said, "If we're lucky, maybe they'll keep us as pets."[366] Or more important, as Ray Kurzweil said, we won't want to make them mad at us.[367] Later in life, Irving J. Good recanted his optimistic view of ultraintelligent machines and concluded they would be potentially disastrous.

Is an ASI Homo nouveau?

There is clearly the notion that a superintelligent machine would treat *Homo sapiens* similar to how *Homo sapiens* currently treat less intelligent species. That's not a good thought for *Homo sapiens*. But I never said my goal was to define *Homo nouveau* as something good; I'm simply trying to define it. So the question is, does a superintelligent computer qualify as *Homo nouveau*?

It certainly doesn't fit anyone's definition of living. It has no metabolism and no compartmentalization to create a fluid environment isolated from its surrounding; that is, it doesn't have cells. First of all, in setting the groundwork for this discussion, I never exactly said that *Homo nouveau* had to be alive according to any particular definition. Second, in my discussion of viruses, which are not alive according to today's definitions, I concluded that there is nothing that would prevent us *Homo sapiens* from declaring a new domain of life called viruses. In the same vein, there is nothing that would prevent us from declaring another new domain of life called superintelligent computers. In fact, maybe the ASI would be the one to declare itself alive! The fact that a superintelligent computer doesn't fit the currently accepted definitions of life seems to be a bit of a technicality to rule it out as *Homo nouveau*.

I said in the groundwork discussion that *Homo nouveau* did need to be a species. A superintelligent computer certainly does not qualify from that point of view either. It is not a separately evolving gene pool because it doesn't have genes. Is this another technicality? Should we change our definition of species? Even if we did, I would hardly think our committees of taxonomists would put a superintelligent computer in the *Homo* genus, or even in the same phylum, kingdom, or domain. Thus, it would never be called *Homo nouveau*.

Although the creation of a superintelligent computer will be a singular

event in our evolution of *Homo sapiens* tools, no matter how it ultimately affects the future and survival of our species, it will not be *Homo nouveau.* I need to look elsewhere.

What about Kurzweil's singularity?

Kurzweil's singularity did not only envision a stand-alone superintelligent computer with superhuman capability. It also described a bionic interface and integration with real human brains using nanobots. It is what Vernor Vinge and others have called IA (intelligence amplification) rather than AI (artificial intelligence). We already are experiencing IA with smartphones and the Internet. It would have taken me years longer to write this same book without IA. Kurzweil's nanobots populating our brains, freely communicating with external computers, and other people will provide unimaginable advantages to those with them. We'll have instant access to all of humanity's information. That information will be digested and summarized or provided in detail, whatever we want. We will be able to upload information-based skills directly into our brains without going through the learning process. We'll also have the equivalent of those Brainets with selected others of our species.

At this point, we are entering the realm of what some call transhumanism. It is the enhancement of the human mental and physical condition using technologies in a manner far exceeding that available through natural and biological processes. Other terms sometimes used to describe this altered state of humanity are posthumanism or extropianism. The resulting improvements in intellectual capacity, physical capacity, longevity, resistance to aging, and other characteristics would be so great that it raises the question whether we should consider these individuals as *Homo nouveau.*

Is such a person a *Homo nouveau*? Is a cyborg with vastly enhanced intelligence because of embedded nanobots a new species? No, at least not as long as the brain nanobots don't change the genome. As long as the brain cyborgs continue to interbreed with everyone else, there will be no cladogenesis. But perhaps that won't be the case. Perhaps only a select few will be able to get the nanobot treatment for whatever reason: cost, discrimination, elitism, religion, or some other reason. Further, what if those in this select group practice a type of behavioral isolation where some or all of them only wish to associate and interbreed with like cyborgs? This would be similar to the discussion of the GenRich in the previous chapter on genetic engineering. As Nicholas Agar said in his essay on transhumanism when referring to a time when humans and posthumans coexist,

> *...it might simply be the case that we find each other so profoundly repellant that interbreeding is mutually unthinkable. We can imagine that this repulsion could be much more profound than that resulting from the racist thinking to which humans seem susceptible, creating reproductive barriers that are more enduring than those racism occasionally creates.*[368]

Then, the sympatric possibilities open up as we have discussed before. Not only could Bill Joy's worst nightmare come true with regard to the social consequences of this technology, but true speciation could happen by any number of genetic chance events that ultimately make these cyborgs reproductively isolated. They would then be *Homo nouveau* coexisting with *Homo sapiens*. There would likely remain some *Homo sapiens* with the nanobot technology that continued to interbreed freely. Nonetheless, because the reproductive isolation of the *Homo nouveau* subgroup would remain, the two *Homo* species would continue to diverge. What the ultimate consequence of this coexistence would be depends on too many factors upon which to even speculate.

Although theoretically possible, this path to *Homo nouveau* seems very unlikely to me. Similar to my thoughts about the GenRich, I don't think behavioral isolation by future enhance brain cyborgs will occur in the first place. As with all technologies that preceded it, this one is likely to become widely distributed in the population over time. There will likely be many variations to the nanobot technology, just as we have lots of different kinds of cell phones, computers, and automobiles. There are also different levels of usage of different technologies, which hasn't led to behavioral isolation in the past. We certainly have techies and nontechies today, and they interbreed.

My conclusion is that a nanobot-enhanced human, even after the singularity, will not become a separately evolving metapopulation and therefore will not be *Homo nouveau*.

These individuals will freely interbreed in the general population. However, as you will see in the next chapter, electronic evolution is too significant to be ignored altogether in this discussion.

What about robots?

A variant of electronic evolution relates to robots and deserves a mention here. Robots are machines that perform tasks that humans or other animals do. They can work independently of humans or under direct human control. Their "brains" are computers. There are many types of robots today that do myriads of tasks in manufacturing processes, warehousing, business services, and home services. Most of the time the repertoire of tasks that any particular robot can perform is limited to specialized and repetitive functions. To call something, or even someone, robotic is to imply a somewhat stilted behavior that is less than human.

That is changing. We are seeing dramatic improvements in robotic technology that is increasingly making robots more capable, more adaptable, more animal-like and even more humanlike. A robot recently hitchhiked across Canada by itself.[369] A robot system called STAR outperformed expert surgeons in reconnecting severed small intestines in pigs.[370] A robot vehicle is delivering pizza in Australia. Robots routinely deliver meals and records in hospitals. They have forever changed manufacturing and warehousing. Some things we call robots don't physically *do* anything; they just "think," talk, or otherwise respond to humans like a search engine, Siri in your iPhone, or Watson in playing Jeopardy. These are sometimes called softbots.* The fact that they have instant access to practically all of human knowledge gives them a bit of an advantage over us. On the other hand, they don't "know" they're doing it. They're unaware that they are robots. Watson didn't go out and celebrate with a beer after its victory. They lack self-awareness. They lack consciousness, whatever that is. Maybe that doesn't matter. We're learning that more and more of what we decide happens subconsciously. What would a humanlike robot really be like? Would they ever have common sense or reflective consciousness? Science fiction portrays robots as superhumans, indistinguishable in appearance but ultra intelligent, ultra strong, ultra everything. Even if we get close to producing such robots, can they become *Homo nouveau?*

There is a tension and a debate as to the future direction of robotics, and it has existed since the early days of robotics. Should robots replace humans or augment humans? Is it AI or IA? John Markoff, the former *New York Times* reporter, reviews this conflict in his book *Machines of Loving Grace.*[371] It is one thing to use machines to replace the mechanics of humans and quite another to replace the human brain. Further, it is one thing to augment the brain and quite another to replace the brain. For example, Google's experiment with self-driving cars seems to be aimed at the AI end of the spectrum—to replace human drivers. However, it may turn out that the best use of this technology may be simply to monitor us human drivers and intervene only when necessary to prevent an accident or other error. That is, keep the human in the loop at all times and use the technology as IA. Or there could be some middle ground where the robot (in this case, the self-driving car) is in control only in some circumstances, such as freeway driving or slow bumper-to-bumper traffic, and the human does the rest. AI and IA may really be two ends of a spectrum and could be quite compatible. However, their design is different, and it depends on the goal and even the philosophy of the designer.

If the goal is to replace the human brain, the first step in the process seems to be to forget trying to program every possible scenario into the brain of a robot. That "brute force" or rules-based method was how IBM's Big Blue beat Garry Kasparov in chess. Instead, the current approach is to try to program a method for the robot to learn from experience, just like a baby does. That is, give their brains plasticity. Actually, to be more precise, the most successful current approaches like Siri and Watson use a combination of rules-based methods to pick off the initial most common scenarios followed by a learning methodology to build out the less common scenarios. This enables the system to begin to be useful to users but then takes advantage of the user experience to improve over time. Second, there is an attempt to build in some form of emotional response—happiness, sadness, fear, and other emotions. These would appear as a childlike smile on something that looks like a child's face, or body language like cowering. They would also need to recognize emotion in

*The term softbot may need to shift meaning. Recently, some robots are being made from entirely soft and flexible materials, including the power source. These rubbery devices are inspired by the octopus and are envisioned for uses that require squeezing through narrow spaces. (See Nature magazine, Vol. 536, 2016, p. 400).

humans—something very subtle and difficult to teach a robot to do. Children do it naturally and are at point A in the Blois funnel. Emotional robots are being designed with the goal to comfort children in hospitals or adults in nursing homes. They would also need to learn tact, which is often a word that means lying or at least not telling the whole truth. That's a really tall order.

In fact, all of this is a really tall order. Softbots are proliferating on the Internet from many companies including Facebook, Google, and Microsoft. Their intent is sometimes simply to be entertaining devices for chatting. Some do targeted functions like scheduling appointments or translating from one language to another. They all use complex neural network software intended to learn from vast troves of recorded conversations and interactions, as well as from the experience in talking with people. In 2016, Microsoft had to remove from the Internet its teenage chat companion, called Tay, after only one day of usage because it began spouting racist and sexist comments that it apparently "learned" from its experience.[372] Was this a sign of success in truly emulating a human, or was it a failure in programming in a superego? It was certainly not the first, and neither will it be the last, bump along the way to AGI.

In the end, this discussion of robots leads to the same problem as discussed previously regarding superintelligent computers. Can we simulate the human brain in the computer brain of a robot or softbot? The only difference is that, for the robot, sometimes we're putting this artificial brain in a mechanical device that has locomotion, sensory inputs, and motor outputs, and sometimes it kind of looks like a person. Although all of these additional robot components are formidable technical tasks, the key issue with regard to robots leading to *Homo nouveau* relates to the robot brain—whether it sits in a robot, in a computer, or in our brain as a nanobot.

Therefore this is the same discussion as occurred regarding stand-alone superintelligent computers. The conclusion is the same. Robots will not be alive, they will not be a species, and they will not be *Homo nouveau*. They will be tools. They could be amazing tools or disastrous tools. On the amazing side, they offer the possibility of an idyllic lifestyle free from boring, repetitive labor and physical risks.[373] They can perform tasks with strength, speed, and agility not possible with our mere human bodies. They are tireless, free from distraction, and mostly error-free. On the disaster side, they threaten anything from massive unemployment[374] [375] to aides in the existential threat from AI. Whatever they turn out to be, they won't be the next in line in our lineage.

In fact, the entire distinction between a superintelligent computer and a robot blurs entirely.[376] The main difference between a robot and a superintelligent computer lies in the mobility of the robot and the various sensors on a robot. The likelihood is that there will be multiple sensors in the environment, mobile and static, connected wirelessly to superintelligent computers, which gives the computers essentially the same or even greater capabilities as robots. The robots simply become extensions or agents of the superintelligent computer. Whether the brain sits in the robot, is stationary in a computer, or is some combination becomes irrelevant.

What is relevant is that neither the most advanced robot nor a superintelligent computer will ever be considered *Homo nouveau.*

I have spent a great deal of time on the subject of electronic evolution. I did not expect to when I began this research. I have included much of this research in the book even though I have concluded that a superintelligent computer, a robot with a superintelligent artificial brain, or even a human with embedded superintelligent nanobots will not be *Homo nouveau.* But as you will see in the next chapter, electronic evolution will play a major role leading to *Homo nouveau,* and regardless of that role, electronic evolution is a game changer for our species.

Chapter 14

The Answers

"How will *Homo sapiens* and *Homo nouveau* coexist?"

The questions are: What comes after *Homo sapiens?* When will that happen, and what will that new species be like? I have spent the past several years educating myself on speciation and looking for clues to the answers. I have tried to think of all the ways this could happen. I have come to my own conclusion. Of course, no one knows the answers. I can only provide my informed speculation.

Speciation by Darwinian evolution has ended for our lineage. We are undergoing what I am calling a revolution in evolution. *Homo nouveau* will not arise from *Homo sapiens* through evolution and natural selection. That is, the manner in which *Homo sapiens* has evolved to this point, going all the way back to the origin of life on earth, will not be the mechanism of our speciation going forward. Note that I am *not* saying that Darwinian evolution has ended for our lineage. In fact, the opposite is true. Darwinian evolution of *Homo sapiens*, as evidenced by changes in our genome, has increased by one hundredfold in the past forty thousand years. Our population has increased in this time by over one thousandfold. We now populate every livable niche on earth's land. We are the epitome of success of Darwinian evolution. In fact, you could say that we are at or even past the "knee" of an exponential rate of increase in Darwinian evolution for our lineage. It took 2.1 billion years to get from the first life on earth to the first eukaryotes, and another 1.2 billion years to get from eukaryotes to mammals. From there it took only 165 million years to get to primates, and another 53 million years to get to the first *Homo* species. From there, it was only 1.8 million years to *Homo sapiens.* At this rate of evolution acceleration, you would think we'd be ready (evolutionarily speaking) for *Homo nouveau* to emerge by Darwinian evolution.

No, Darwinian evolution has not ended for our lineage. It has accelerated since *Homo sapiens* has come on the scene. But *speciation* by Darwinian evolution has ended. With a hundredfold increase in the rate of genetic change, one might expect that natural selection would lead to cladogenesis of *Homo nouveau* in only ten thousand years. That would be an astoundingly short period of time. But another "knee" has already begun to usurp that timetable: our exponential growth in intelligence. This exponential growth is the result of a confluence of science and technology advances that is enabling our ability to alter an evolutionary process billions of years old. The impact of this alteration is twofold.

1. Ironically, it has already begun to slow the recent increase in rate of Darwinian evolution of humans. The essence of Darwinian evolution is that people with advantageous genetic mutations tend to have more babies, and that people with disadvantageous genetic mutations tend to have fewer babies. That was true in humans until the past two hundred or so years, when infant mortality was an order of magnitude higher than it is today and a large proportion of the population died before reaching childbearing age. Our advances in hygiene, sewage management, infection control, nutrition, and medical capability have dramatically reduced the effect of the selection component of Darwinian evolution in our lineage in the most recent two hundred years.[377] Simply put, we keep a lot more people alive long enough

to have children than we used to. That has dramatically reduced the impact of negative selection. Couple this with the facile interbreeding of our species everywhere on earth, and you have a recipe for greatly inhibited speciation by natural selection.

2. More significantly, our intelligence is enabling another method of speciation to take over. This revolution in evolution will lead to *Homo nouveau* far faster than historical Darwinian evolution.

I have also rejected several other possible mechanisms as the most likely factors leading to *Homo nouveau.* It will not be the result of a catastrophic event, and neither will it be the direct result of electronic evolution—although as you will see, the latter will have an indirect effect. *My conclusion is that we will genetically engineer Homo nouveau.* Remember, it takes two factors to make a new species: genetic change and a barrier to interbreeding with the ancestor species. So how do I envision that happening by genetic engineering?

To illustrate my vision, I'm going to take you to an imaginary time 350 years into the future to demonstrate the impact of this intelligence growth on our speciation. That's a blink of an eye, evolutionarily speaking. Figure 23 says it all.

FIGURE 23

Donald Simborg$_{14}$ *Homo sapiens*

Cyborg Simborg *Homo nouveau*

The person on the left is Donald Simborg$_{14}$. (This name notation using subscripts came in use in the twenty-second century and replaced the older naming notation: Donald Simborg, XIV). Donald $Simborg_{14}$ is a *Homo sapiens* and a descendant of your author. The person on the right is Cyborg Simborg, a *Homo nouveau*. They are similar in appearance (except for the eyeglasses). They are related in that they share a common ancestor five generations before Donald $Simborg_{14}$. Cyborg Simborg is a singularitarian with a nanobot-enhanced brain. The singularity, as defined by Kurzweil, was achieved in 2145, exactly one century later than predicted by Kurzweil. (By the time of this photo, the term singularitarian shifted in meaning from someone who believes that the singularity will occur to someone who elects to embed nanobots in his or her brain.) If I were to do a genome comparison of the two, 100 percent of their coding genes would be the same. At the nucleotide level, their genomes would differ by only a few thousand out of the three billion nucleotides in each of their genomes. That is because five generations have passed between Donald $Simborg_{14}$ and their shared ancestor, Donald $Simborg_{9}$, who was Cyborg Simborg's father. During that time, this relatively small number of nucleotide differences was introduced by mutation and meiotic mixing of alleles from the maternal line of Donald $Simborg_{14}$. I know this is confusing, but bear with me, and it will become clear. Figure 24 below shows the detailed timeline for these events.

In 2273, the Science and Genomic Administration (formerly the Food and Drug Administration) approved germline genetic engineering for clinical usage. This was well after the singularity, and our knowledge of genetic engineering was greatly accelerated by it. One problem with all research prior to the singularity was that the pace of discovery and publications greatly exceeded anyone's ability to absorb it. This made research fragmented, redundant, costly, and inefficient. Nanobot technology and the singularity eliminated that problem in singularitarians. It accelerated the pace of research in germline genetic engineering and quickly led to our greatly increased ability to perform it, our belief that it would be safe, and hence its approval by the Science and Genomic Administration.

In 2200, Ethan Simborg, the brother of Donald $Simborg_{10}$, was born. He subsequently became a singularitarian in 2226 and changed his name to Cyborg Simborg. In 2277, Cyborg Simborg, at the age of seventy-seven, underwent genetic surgery to eliminate cellular senescence. In fact, over a million *Homo sapiens* underwent the same genetic surgery during that era. This particular form of genetic surgery was the most popular form of genetic engineering at that time. Since aging is a complicated process involving many different types of somatic cells, the DNA cutting tool and vector used was developed to target virtually all cellular types including germline cells. Thus, the procedure did not have to be repeated in the offspring. The genetic engineering was done by other singularitarians.

Yes, with the singularity, our knowledge of genetic engineering had advanced exponentially to the point where it became as routine as zapping cancer cells with CRISPR-armed nanobots. That is, it was happening every day by genetic surgeons. It was later determined that our mastery of genetic engineering at that point in time was not perfect. We were very good at it, but not perfect; our genome and epigenome are too complicated for that. Although we had completely simulated the human brain by that time, we had not yet completely simulated the cellular metabolism of a single cell. There was a sort of Clockwork Orange character to the early years after the approval of germline therapy in the United States, where unexpected and even strange things happened. One of those unexpected things was off-target mutations as a side effect—the same problem that had plagued the most early experiments in genetic engineering.

Although by 2273 we had a complete understanding of the human brain, including its genetics, we had less of an understanding of the aging process. The Human Anti-aging Institute, first proposed in the late twenty-first century, did not ever get fully funded. There was too much concern that the population was already too high, and this would only make that situation worse. Nonetheless, genetic surgeons went where the money was. Living longer and healthier was more important than becoming smarter to many people who could afford the concierge genetic surgeons. Although the Methuselarity was ultimately achieved, it was not without problems. Aging is complicated and involves many different genes and epigenetic processes, thus making the genetic engineering more risky. The result was likely an off-target mutation of an epigenetic process that affected everyone undergoing the genetic surgery to eliminate cellular senescence.

The effect of this off-target mutation is that when Cyborg Simborg and other *Homo nouveau* species mate with any *Homo sapiens*, the fetus miscarries. When a male and female *Homo nouveau* mate with each other, the fetus develops normally, and the offspring are completely normal. Thus *Homo sapiens* and *Homo nouveau* are separately evolving metapopulations with a barrier to successful interbreeding. The exact cause of this postzygotic reproductive isolation is not known. Normally, there are mechanisms during pregnancy that prevents the mother from rejecting the foreign proteins related to the father's component of the fetus. In the hybrid case involving *Homo nouveau* and *Homo sapiens*, that mechanism no longer works and spontaneous abortion (miscarriage) results. It's as simple as that, and *bang*—a new species is created. Both the genetic changes regarding cellular senescence and the barrier to interbreeding with *Homo sapiens* were somehow introduced by the same genetic manipulation. Not quite the big bang of the other kind of singularity, but still pretty remarkable.

The photo above was taken on March 27, 2365, during a family reunion. Donald $Simborg_{14}$ is in the line of *Homo sapiens* who chose not to become singularitarians. He was born in 2290 and was 75 years old at the time of the picture. Cyborg Simborg was 165 years old at the time of the picture. Cyborg Simborg wears glasses because the genetic engineering to eliminate cellular senescence is incomplete with respect to presbyopia (far-sightedness). Figure 24 shows the timeline of events leading to this photo.

Figure 24

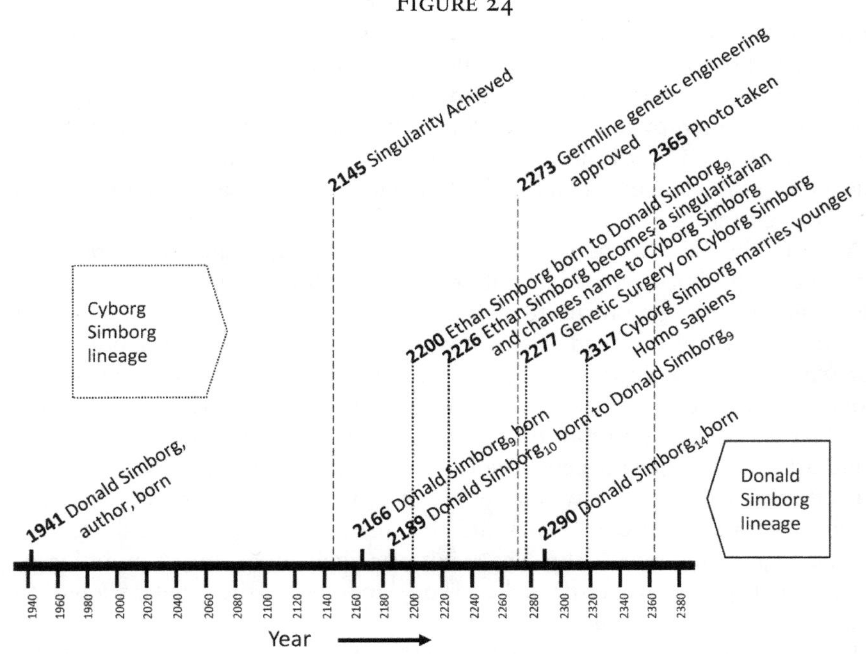

As you can see from figure 24, Donald Simborg$_{10}$ and Cyborg Simborg are brothers. However, Cyborg Simborg (born Ethan Simborg) became a *Homo nouveau* after undergoing genetic surgery in 2277. He subsequently married a younger *Homo sapiens*, but several subsequent pregnancies of his wife failed by miscarriage.

Let me be clear. The singularity is not necessary for this path to *Homo nouveau*. The cause of the speciation event was the genetic engineering and its unanticipated off-target consequence likely regarding an epigenetic process. This would happen without a singularity as well, but it would happen much later. It is the same scenario as depicted in figure 21. There will be singularitarians among both *Homo sapiens* and *Homo nouveau*. That is not what distinguishes the two species.

The particular genetic engineering on-target change that led to this off-target speciation event was related to aging, not to a brain enhancement or some other better understood genetic engineering area. This is not a particularly relevant point because it could have been many other on-target genetic engineering changes that led to the unanticipated off-target mutation leading to the postzygotic reproductive isolation of *Homo nouveau*.

If you had visions of *Homo nouveau* as large-headed creatures with very high foreheads and weak muscles that communicated silently, I'm sorry to disappoint you. Perhaps one day *Homo nouveau* will evolve into that, but long after speciation from *Homo sapiens* has occurred. Once the two *Homo* metapopulations begin evolving their lineages separately, anything can and will happen. Mutations and meiotic mixing will continue to occur, natural selection will continue to operate slowly, and genetic engineering will continue to be performed. My prediction is that genetic engineering will be the driver of all future speciation events in our lineage, not natural selection. Because of the singularity, our knowledge and capability of this technique will continue to grow exponentially, and new species in our lineage will be created, both intentionally and accidentally, on much shorter timeframes than previously.

When did speciation of *Homo nouveau* occur? Sometime in the late twenty-third century. The exact timing is not precise because it took thousands of genetic surgeries and several generations to establish the separately evolving metapopulation. Nonetheless, that's a very short time for speciation to have occurred. *Homo sapiens* has fundamentally and irreversibly altered evolution. By the early twenty-first century, we had already altered the evolution of thousands of species of plants, animals, and other living things by genetic engineering. We continued to do so. Natural selection–induced changes were relegated to SCs in the de Queiroz diagram. Some natural selection changes will continue to lead to cladogenesis in some species. Those will be few and far between compared to the genetically engineered events.

We are a eukaryote. All other plants and animals are also eukaryotes. Genetic engineering will not change that. Some of the other eukaryotes will continue to be altered by genetic engineering performed by either *Homo sapiens* or *Homo nouveau*. The rest of the eukaryotes—the ones we're not interested in changing by genetic engineering—will also be dramatically affected by the *Homo* species, and generally not in very good ways because of the destructive effects *Homo* has on earth. Nonetheless, they will all continue to evolve by natural selection. The least affected species by *Homo* will be in the other two domains, archaea and bacteria. They'll continue to evolve, adapt, and dominate the world in numbers and biomass as they always have. They will continue to constitute most of the DNA that is in or on our bodies.

How will *Homo sapiens* and *Homo nouveau* coexist? No differently than various diverse groups of *Homo sapiens* coexist today, at least for millennia. It hasn't been very long (only thirty thousand years) since we've had two or more human species existing at the same time. In fact, that has been the usual circumstance. But this time it will be different. We won't be geographically separated. There will be no allopatric speciation. Our contact won't be limited to the borders or front wave of migrations as it was between the early *Homo sapiens* and Neanderthals. We'll be much more alike and integrated from the moment of speciation. There will be *Homo nouveau* that will be Asian, Hispanic, African, and everything else we have today. The two human species will go to school and work together, live in the same neighborhoods, speak the same languages, and eat the same foods. They will intermate freely but still remain separately evolving metapopulations owing to their postzygotic reproductive isolation.

Eventually the two species will diverge because of random mutations, meiotic recombination being acted upon by natural selection, and further genetic engineering. But long before ten thousand years go by to allow speciation by natural selection, there will likely be other *Homo* species created by genetic engineering of either *Homo sapiens* or *Homo nouveau*. Speciation of our lineage has been changed forever.

It is quite likely that one or the other of the two human species will die out. Which one and how that will happen is anyone's guess. There are too many random events that will happen to predict it. Both species will be threatened by the possibility of an artificial superintelligence eliminating them. This could even happen before the speciation of *Homo nouveau*. Assuming we will manage to control the AI phenomenon and the ASI will be friendly, the two *Homo* species will continue to diverge both by natural selection and further genetic engineering. This will most likely be related to enhanced brain functions. One species will likely become superior to the other in terms of survival capability or the ability to conduct warfare. Bill Joy's worst fears could come true. But that's pure speculation, as is everything else in this chapter.

Afterword

"We probably have maxed out on the survival benefits of increasing intelligence...What we do know is that the brain is both amazing and probably sows the seeds of our destruction."

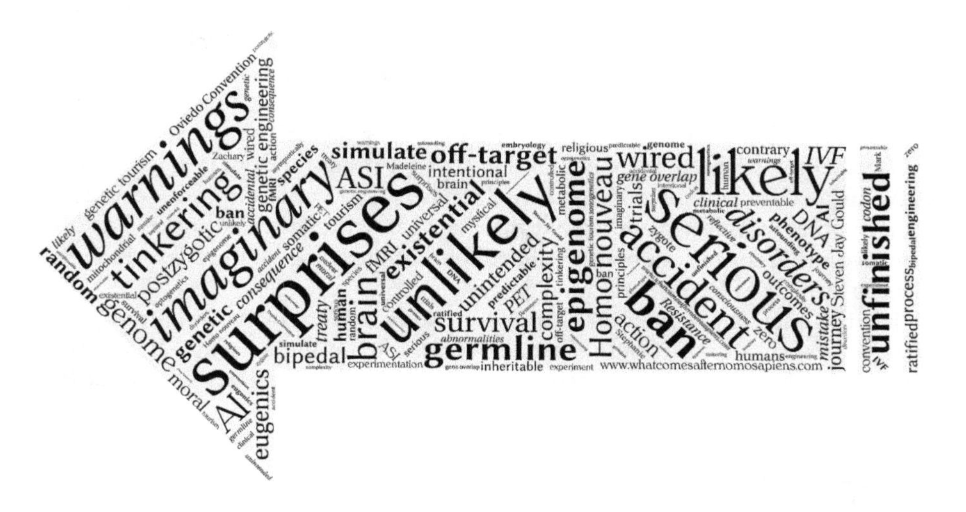

Although the particular scenario described in chapter 14 is imaginary, my conclusion regarding the most likely mechanism leading to speciation of *Homo sapiens* into *Homo nouveau* is based on serious study. In my opinion, that mechanism will be genetic engineering by *Homo sapiens*. Specifically, it will be an off-target epigenomic change causing postzygotic reproductive isolation of *Homo nouveau* secondary to some popular genetic enhancement. Although the particular scenario described in chapter 14 is unlikely, one of the many possible similar scenarios seems very likely to me; in fact much more likely than the other possible mechanisms that I researched.

As I said in the preface, there were many surprises for me in my journey to the answers. Perhaps the biggest surprise of all is that the answers really are the result of an accident. They are the unintended consequence of a *Homo sapiens* act. It is not some deterministic or logical conclusion to a predictable process. On the other hand, why should that surprise me? All of evolution is an accident, a random set of events that, but for other random events, could have turned out entirely differently. As Stephen Jay Gould said, if we could find another earthlike planet, plant the exact same organic seeds on it, and allow evolution to start all over, the likelihood of anything remotely resembling *Homo sapiens* evolving is miniscule.[378]

But this accident is different. The *Homo nouveau* that I envision is not a random event of nature, but rather an accident in a different sense. It is the unintended consequence of an intentional action by *Homo sapiens*. It is a human mistake. Could a mistake like this have been prevented by a ban on germline genetic engineering as called for by the Oviedo Convention and numerous other national and international bodies? Germline genetic engineering raises the ethical question of risking serious consequences on unborn humans without their consent. It raises the ugly scepter of eugenics. It is contrary to religious and moral principles of many, if not most, living humans today.

My view is that there is little possibility that such a universal ban will ever happen, and even if most countries ratified such a treaty or convention, it would be unenforceable. Resistance would break down over time, as it has for IVF, PGD, somatic genetic engineering in humans, and mitochondrial germline engineering. The benefits are too compelling; inheritable genetic disorders are too terrible. The resistance will break down gradually, first allowing very controlled experimentation in some countries, and then limited clinical trials. Our ability to cure at least some disorders using germline engineering will be demonstrated somewhere, sometime. That country will then approve its use. Genetic engineering tourism will occur. We already have at least one such example where a physician at a fertility clinic in New York City took his patient's and donor's unfertilized eggs to Mexico, performed mitochondrial germline editing (as described in appendix 5) where it is legal to do so, in order to avoid the United States ban on this procedure.[379] Sooner or later—in my view, later—germline genetic engineering on nuclear DNA will be done largely successfully in most countries. A certain amount of complacency will set in. Accidents will happen, as they do today in almost all medical procedures.

It is the complexity of the human genome that creates the likelihood of accidental outcomes when tinkered with by *Homo sapiens*. The genome is the most amazing thing in the universe. That a linear string of four chemical compounds sets in place the beautiful and magnificent symphony of embryology is beyond mystical. How a single cell, a zygote, divides and differentiates into a fully functional and complex organism under the control of sequential epigenomic influences is profound beyond words,* At the same time, it is so variable and fraught with error that it is amazing that it works as well as it does. Certainly it is far from perfect. We have tragic genetic defects, myriads of cancers, and metabolic abnormalities to plague human and all other species' existence. Our genome is filled with the remnants of previous species in our lineage going all the way back to ancient bacteria and viruses. Now add to that mix the human tinkering of genetic engineering, and it is no wonder that unexpected things happen. We should know better, but we can't help ourselves. Our brains are wired to experiment.

And what about that brain? What exactly distinguishes our brains from other species? What is our unique brain phenotype? Is it common sense? Is it reflective consciousness or shared intentionality? Is it the way we represent language mentally or express it with fictive language? We don't know all the answers. What we do know is that the brain is both amazing and probably sows the seeds of our destruction. The amazing part is how it allows us to think, to have common sense, to have reflective consciousness and shared intentionality, and to create fantastic tools. Tools like fMRI, PET, optogenetics, and sonogenetics are allowing us to visualize how the brain is connected and how it functions to some extent. But seeing how it is wired is not the same as understanding how we think. We haven't invented the right tool for that yet, but we will; I'm convinced of that. When we do, we'll able to simulate the brain in another amazing tool, the computer. I'm afraid that will be our downfall. As I concluded at the end of chapter 11, we probably have maxed out on the survival benefits of increasing intelligence. I defer to greater brains than my own who are telling us that artificial intelligence will be the end of *Homo sapiens* or any other *Homo* that follows: Bill Joy, Stephen Hawking, Vernor Vinge, Shane Legg, Stuart Russell, Max Tegmark, Nick Bostrom, James Barrat, Michael Anissimov, Elon Musk, and Irving Good. Brilliant minds, Nobel Prize winners, renowned inventors, and IT pioneers—all giving us warnings!

One surprise in my research was our puzzlement about why we became exclusively bipedal. I should rephrase the question. Anything that happens in evolution is random. Why it was selected to persist is the correct question. Some people have said that walking upright is what makes us human. Hardly. It is a minor trait compared to our brain. I've reviewed about a half dozen or so theories as to why we walk upright in chapter 7, but the result is less than satisfying or convincing. It seems to me that upright posture has caused more problems than it solved, which flies in the face of natural selection. Clearly it came before our brains got big, so perhaps bipedalism wouldn't have happened had the sequence been reversed. And no one ever said that evolution

*For a thorough and beautifully written description of human embryology, see Jamie A. Davies's book Life Unfolding: How the Human Body Creates Itself.

always led to the best results in the long run.

Another surprise is that we still haven't even agreed on what a species is. I've now read about five different books on the subject, and each one seems to be more difficult to understand than the other. This topic is more than one of John Hands's deep canyons—it's more like a black hole. In the end, I'm convinced there is no answer to the species problem because I believe that species are not real entities but simply concepts in the mind of *Homo sapiens*. Arguing about it will not solve the problem. The definition is basically arbitrary. *Homo nouveau* will be a different species from *Homo sapiens* only because it fits my adopted definition that it will be a separately evolving metapopulation and remain so because of its barrier to successful interbreeding.

Even more basic is that we can't agree on the definition of life. Whatever it is, we also don't understand how it began on earth or anywhere else. There are lots of theories about lightning bolts, UV radiation and other energy sources interacting with some kind of primordial chemical soup in the sea or puddles or rock surfaces or hydrothermal vents. Maybe life arrived on a meteorite, but that simply moves the question elsewhere. If we knew the answer, we could reproduce life in a laboratory. We can't. Even evolution doesn't seem to have the answer since, as far as we know, new life has not emerged again since the original event.

It is remarkable how similar our genes are to more primitive species, even going back to the most primitive forms. It didn't take much to evolve from an ape to a human, and even less to evolve from our predecessor human, whatever that was. In the hominid lineage, we seem to be asymptotically approaching zero for the number of gene differences needed to create a new species. That is why I feel comfortable with projecting *Homo nouveau* as having 100 percent gene overlap with *Homo sapiens*. By gene, I'm referring to the common definition as a DNA sequence that leads to the coding of a protein or a coding gene. We now know that there are many more noncoding genes that generate the epigenome. It is in the epigenome where most of the differences in *Homo* species exist, and that is certainly true for *Homo nouveau*.

My concept and understanding of the genome has undergone an evolution of its own. When I started writing this book, I was still stuck at Mendel's peas, big B and little b, genes make proteins, and all the things I learned from high school through medical school. Looking back, even as a physician, I was still in pre-school when it came to my understanding of the genome. I probably still am. The heart of the revolution in evolution is the epigenome. It is the ultimate exaptation force: repurposing the same set of DNA sequences for different uses. What mainly changed in going from Ardi to Lucy to *Homo habilis* to *Homo erectus* to *Homo heidelbergensis* to *Homos neanderthalensis* and *sapiens* and finally to *Homo nouveau* was not our genes, but our epigenome.

There are many unknowns in the lineage of *Homo sapiens*. There are many controversies regarding which species, exactly, are in our lineage, and which one is our immediate predecessor, and when exactly *Homo sapiens* emerged. I spent a lot of time in my research trying to sort through those controversies, largely without success. We simply don't know. In retrospect, it didn't matter regarding the answers. If and when we ever figure out the answers to those lineage questions, they will not alter my conclusions. However we got here and whenever we got here, we are here, and we are what we are. We know what our genome is, and we know how to alter it. In the end, that's what really mattered in order to reach my conclusions.

Homo nouveau will not be as astounding an event as I had imagined when I began this research. At first, it will be barely noticeable, if at all. Over time, our advanced tools will surely recognize the new species among us, and perhaps we will even attempt to alter the postzygotic barrier by genetic engineering. If we are successful at that reengineering, *Homo nouveau* will be short lived. More likely we will not be successful, and the two species will continue to evolve by both natural selection and genetic engineering, including the changes introduced in our attempt to reengineer the speciation event.

The end result of our genetic tinkering will likely be multiple *Homo* species. By that time, new definitions of species will be created by one or more of the *Homo* species. The species problem will be compounded by different definitions of species by different species. Won't that be an interesting problem? It will take AI to resolve it one way or another—perhaps by eliminating them all.

And speaking of AI, one can reasonably argue that chapter 14 should be entirely rewritten. I concluded in chapter 13 that electronic evolution and robotics are likely to profoundly change the course of human evolution but that they in themselves would not lead to *Homo nouveau*. That was because whatever the ASI turns out to be, it would not fit any current definition of species—certainly not one in the *Homo* lineage, and not even in the Eukarya domain. It doesn't even fit anyone's definition of life. But that's a *Homo sapiens* perspective. An ASI may define species, domains, and life entirely differently, and I would speculate it would consider itself just as alive as we are. Who will judge whose view is correct? By my projection, *Homo nouveau* won't emerge for at least another couple of centuries. Most experts expect ASI to be achieved prior to that. I am predicting that also. If so, even if *Homo sapiens* survives beyond that point, it can be argued that what comes after *Homo sapiens* is whatever the ASI defines as itself.

Yuval Noah Harari in his book *Sapiens*, talks about a new meaning of the phrase "intelligent design." Normally this refers to the creation of humans and all other living creatures as works of a divine entity or God. Harari suggests that the meaning of this term could be transformed to describe the abilities of *Homo sapiens* to alter and even create new life. Is my prediction of *Homo nouveau* an example of this new meaning of intelligent design? Perhaps it would be better characterized as unintelligent design. After all, it is an unintended consequence of our actions. And the notion that we have set in motion the creation of an ASI seems the most unintelligent idea imaginable.

This book is certainly not finished, and the answers are certainly not resolved. Not a week goes by that I don't read something newly published that is relevant to the answers. Genetic engineering, my conclusion regarding the mechanism for the creation of *Homo nouveau*, is advancing at an astounding pace. We now have the ability to change the genetic code (the codon) of organisms so that we could introduce other amino acids into proteins beyond the twenty that have been used for billions of years. This would enable us to create totally new and synthetic types of organisms with incredible features. Even more remarkable, we can now add two additional unnatural nucleotides to the genome of *E. coli* so that it has six nucleotides instead of four to use to code proteins. We are changing a genetic coding system that has been stable in nature for almost four billion years! Incredible. Maybe *Homo nouveau* will appear even earlier than I'm predicting. Maybe, the nature of *Homo nouveau* is totally beyond anything I can now imagine. At some point, I had to stop writing this book and publish it. I do not intend that to be the end, however. I invite knowledgeable experts, and others like me who are not experts, to continue this discussion on my website, www.whatcomesafterhomosapiens.com.

Acknowledgments

Homo sapiens tools: without them, I could not possibly have written this book. I used PubMed, Google Scholar, online access to scientific journals, Bing, Wikipedia, and countless Internet searches, e-mails, and web pages. It was all IA. Much of what is in Wikipedia and the Internet is inaccurate. I took very little from a single source without verifying it elsewhere. Whenever possible, I went to the original source publication for information. Peer-reviewed journal articles were my primary source, as were numerous texts and full-length books from experts in the various fields. That is not to say there isn't inaccurate material there too, but it is the best I could do.

Homo sapiens. For the time being, we're still the smartest species on earth, and I made as much use of them as I could. My wife, Madeleine Simborg, was the first and most valuable reviewer. She is not a reader of nonfiction and generally doesn't enjoy it. She didn't enjoy the first version of this book. She said it would never sell like Atul Gawande's *Being Mortal*. She said it was dense and boring. She said it needed massive editing and rewriting. All of that provided me with much-needed guidance, and it spared you, the reader, from the earlier versions. Unfortunately, perhaps, it did not spare you from the final version.

My second reviewer, Bob Berman, is an avid reader of nonfiction and basically panned the first version, too. He said there were too many lengthy and uninteresting irrelevant parts, especially about species, speciation, genetic engineering, and artificial intelligence. The final conclusion was not understandable. He failed to see the connection between that first version of the electronic evolution discussion and the answers. Again, Bob's critique led to many revisions. I've moved a lot of material that was less relevant into appendices. However, readers may still think I have not cured all the problems he identified.

Barbara Bonfigli, an author herself, gave me valuable advice on writing style—a problem likely still apparent. She also suggested the name *Homo nouveau* for the possible future species. Likewise, my long-time friend and accomplished writer and columnist Penny Preston gave me guidance on navigating the publication process. The continual coaching from Barbara and Penny led to significant improvements in the graphics and other aspects of the book.

Gregor Ehrlich raised a point that has plagued all versions of the book: Is this meant for an audience already somewhat versed in some of the sciences and technologies involved—or not? Is it a textbook or a book intended for broad readership? Are the deeper dives into some of the subjects really necessary? If it is meant for scientists, much of the book is unnecessary. If it is meant for average readers who are not scientists (which it is), then much of the book is not easily readable.

On it went, through many reviewers and many revisions. Perhaps most telling during the early drafts was the number of friends who expressed interest in reading my manuscript, only to never return any comments after I sent them drafts. I guess they wanted to remain friends. I am particularly grateful to Pam Nicholls and Terry Colyer, great friends who, although admitting they couldn't quite get through the book, nonetheless gave me some great feedback, particularly about the website and how it could be better integrated with the book. My daughter, Stephanie, provided her artistic eye to improve the website design, and my son, Mark, a writer, greatly streamlined the language of the website and a portion of the book.

I quickly learned that sending out complete drafts of my revised book wasn't the best way to get feedback. That was particularly true of some of the writers of authoritative books on related subjects. They generally declined to review the book, and it was somewhat comforting to know that their lack of comment reflected the fact that they had *not* read the book.

One tool that was effective with the authoritative sources was to ask specific and pointed questions. This turned out to be extremely helpful in clarifying some of the many points that were confusing to me as a nonexpert in their fields. I am particularly grateful to Chris Stringer, Nessa Carey, Liram Carmel, Paul Renne, Brian Crother, Ann Gibbons, Svante Pääbo, Matthieu Landon, C. Owen Lovejoy, Aubrey de Grey and Eugene Harris in that regard. I owe a special thanks to Kevin de Queiroz for creating his useful definition and diagram on speciation, which I have borrowed liberally, and for engaging with me in a stimulating dialogue via e-mail about the reality of species.

Since, as they say, luck favors the prepared mind, I must mention the stroke of luck I had in happening upon Eleanor Johnson of Weymouth, Nova Scotia. She is a graphic designer whom I originally contacted to convert my images to a higher resolution suitable for publication. In the process, she discreetly suggested that she could "clean up" some of my figures. The beautiful renderings you see in the book have only vague resemblance to the originals I produced and I will be forever grateful to her for that. She also produced all of the wonderful chapter title images and the book cover. My thanks for one of the figures, however, goes to my grandson, Zachary Breitbard, who taught me how to create figure 23.

Although still flawed, this published version is a different species from the original.

Appendices

Appendix 1

Table of Species Definitions

(From Kevin de Queiroz, "Species Concepts and Species Delimitation," *Systematic Biology* 56 (6): 879–886, 2007)

Alternative contemporary species concepts (i.e., major classes of contemporary species definitions) and the properties upon which they are based (modified from de Queiroz, 2005). Properties (or the converses of properties) that represent thresholds crossed by diverging lineages and that are commonly viewed as necessary properties of species are marked with an asterisk (*). Note that under the proposal for unification described in this book, the various ideas summarized in this table would no longer be considered distinct species concepts (see de Queiroz, 1998, for an alternative terminology). All of these ideas conform to a single general concept under which species are equated with separately evolving metapopulation lineages, and many of the properties (*) are more appropriately interpreted as operational criteria (lines of evidence) relevant to assessing lineage separation.

Species concept	Properties
Biological	Interbreeding (natural reproduction resulting in viable and fertile offspring). Advocates/Reference: Wright (1940); Mayr (1942); Dobzhansky (1950). Mayr (1942); Dobzhansky (1970).
Isolation	*Intrinsic reproductive isolation (absence of interbreeding between heterospecific organisms based on intrinsic properties, as opposed to extrinsic [geographic] barriers). Advocates/Reference: Paterson (1985); Masters et al. (1987); Lambert and Spencer (1995)
Recognition	*Shared specific mate recognition or fertilization system (mechanisms by which conspecific organisms, or their gametes, recognize one another for mating and fertilization). Advocates/Reference: Van Valen (1976); Andersson (1990)
Ecological	*Same niche or adaptive zone (all components of the environment with which conspecific organisms interact). Advocates/Reference: Simpson (1951); Wiley (1978); Mayden (1997)
Evolutionary (some interpretations).	Unique evolutionary role, tendencies, and historical fate Advocates/Reference: Grismer (1999, 2001)
Cohesion	Phenotypic cohesion (genetic or demographic exchangeability) Advocates/Reference: Templeton (1989, 1998a)

Phylogenetic	Heterogeneous (see next four entries)
Hennigian	Ancestor becomes extinct when lineage splits Advocates/Reference: Hennig (1966); Ridley (1989); Meier and Willmann (2000)
Monophyletic	*Monophyly (consisting of an ancestor and all of its descendants; commonly inferred from possession of shared derived character states). Advocates/Reference: Rosen (1979); Donoghue (1985); Mishler (1985)
Genealogical	*Exclusive coalescence of alleles (all alleles of a given gene are descended from a common ancestral allele not shared with those of other species). Advocates/Reference: Baum and Shaw (1995); see also Avise and Ball (1990)
Diagnosable	*Diagnosability (qualitative, fixed difference) Advocates/Reference: Nelson and Platnick (1981); Cracraft (1983); Nixon and Wheeler (1990)
Phenetic	*Form a phenetic cluster (quantitative difference) Advocates/Reference: Michener (1970); Sokal and Crovello (1970); Sneath and Sokal (1973)
Genotypic cluster (definition)	*Form a genotypic cluster (deficits of genetic intermediates; e.g., heterozygotes). Advocates/Reference: Mallet (1995)

Appendix 2

Are Species Real?

There is a fundamental question that is still being debated among the experts in taxonomy and biology. Are species real, or "are they simply theoretical constructs of the human mind?"[380] By real, I mean is there some existent set of species in nature that is waiting to be discovered and described by us? Or alternatively, is there some existent set of organisms in nature that are waiting to be discovered, described, and then organized in our brains into something we call species? Keep in mind that not every living animal is in a species. For example, a mule, which is a hybrid of a female horse and a male donkey, is sterile and therefore is not considered a species. It doesn't reproduce, doesn't have its own genetic lineage, and doesn't fit any of the definitions in appendix 1.[381]

Coyne and Orr, in their seminal book *Speciation*, go to some length to argue that species are real, at least in the sexually reproducing part of the taxonomy. After providing two rather unconvincing (even to themselves) arguments based on common sense and the concordance of folk and scientific classifications, they base their argument on the fact that modern methods of statistical cluster analysis corroborate the notion of real species. They say, "It is important to realize that these methods cannot identify such groups if they do not exist."[382] I personally find their argument weak and unconvincing. The statistical techniques do certainly define discreet clusters of parameters related to species, but not always. There are gray areas and mathematical limits that are arbitrarily set, sometimes in order to define a cluster. Even the concept of a cluster is a construct of the human mind. The exact same cluster analytical techniques can be applied to marketing research, religious beliefs, political surveys, and just about anything about which data can be collected. It is tautological to say that any groups so defined by these techniques are real.

I contrast the notion of species to our knowledge of the set of elementary particles that make up matter, or the chemical elements that make up the periodic table. A chemical element is defined based on the number of protons in its nucleus. We do not have committees deciding whether potassium is an element, or how many protons are in a potassium nucleus. Once *Homo sapiens* conceived of the definition of chemical elements, it was only a matter of time before we discovered all of them—and once discovered, they don't change. The same is true of fundamental particles that make up the atom. Committees don't decide whether neutrons and protons are particles in the atomic nucleus. Although there is still uncertainty and even some debate regarding the exact makeup of an atom at the subatomic level, and the model of the atom is still a theory, with better tools and observation, the existent reality becomes better known. There is no existent reality when it comes to species. We can define them simply with our brains any way we choose.

Continuing with the chemical element analogy, the periodic table is more like a taxonomy. It is a way of organizing the elements according to some set of rules—in this case, the atomic number and the chemical properties of the element. The periodic table is a concept created by *Homo sapiens*. There is more than one way to do this. We did not discover the periodic table; we created it. The periodic table allows logical groupings or classification of elements. These groupings, such as the halogens or the noble gasses, would correspond roughly to a genus. Potassium would then correspond to a species, and the isotopes of potassium would correspond to subspecies. That would be one way to classify the elements. There could be other ways to organize the elements that would create different groupings. For example, we could group them according to whether they occur in nature as solids, liquids, or gasses at some arbitrary temperature and pressure.

Fundamental particles make up atoms. Atoms make up the elements. Elements make compounds. Some compounds are carbon-based and are called organic compounds. Complex organic compounds can take the form of amino acids. Amino acids make up proteins. Other complex organic compounds make nucleotides. Nucleotides make up RNA and DNA. RNA and DNA determine the proteins that constitute entities that can replicate themselves, like viruses and cells. Cells organize into living entities we call organisms. (See chapter 13 for a discussion of the meaning of living.) The possible combinations of nucleotides that make up RNA and DNA are astronomical in number and have led to millions of entities that are living organisms today. Up to this point, these are all objective realities that have been discovered using the tools developed by this remarkable species, *Homo sapiens*. It is analogous to determining that chemical elements consist of nuclei of varying numbers of protons.

It is when we take the next step (which we call taxonomy) and try to sort out these millions of living entities into some logical order that we leave the realm of objective reality and enter the realm of concepts that exist only in the brains of *Homo sapiens*. Species are as ephemeral as the tools used to define them or the committees it takes to declare them. Just look at the history of the yellow-rumped warbler or giraffe as described in chapter 3. *Homo nouveau* will be simply a concept, and our goal is to agree on a definition that is consistent with our existing knowledge and then determine when that definition is met to declare the new species. The same is true of the many other organisms that exist today but have not yet been discovered. It is estimated that we have discovered and categorized only about 20 percent of the Eukaryote species that exist today.[383] At the rate we are discovering new organisms and categorizing them into species, it is estimated that it will take four hundred years to develop a list of species that covers every living organism.[384] But once organisms are discovered, their names and the species to which they belong are concepts only in our brains. They are not existent realities, like chemical elements. As the world-renown paleoanthropologist Chris Stringer states, "species concepts ... are, after all, humanly created approximations of reality in the natural world."[385]

Kevin de Queiroz (see chapter 3) is among those who insist that species are real. After some communication with him on this issue, it is clear to me that our differences revolve around the meaning of the word real rather than the meaning of the word species. To me, real or existent reality relates to things we can validate or know by means of one or more of our five senses of observation, or by using tools that extend our senses of observation. Things that are not real are things we can know only by describing them with our fictive language (see chapter 11 for a discussion of fictive language). Although we also use our fictive language to talk about real things, it is not a necessary requirement in understanding something real. Kevin de Queiroz states, "By 'real' I just mean something that exists independent of human perceptions, conceptions, etc." We have agreed to disagree.

This all reminds me of the baseball umpire joke: There are subjective umpires who say, "If I think it's a strike, I call it a strike." There are objective umpires who say, "If it's a strike, I call it a strike." And then there are existentialist umpires who say, "They ain't nothing until I call them." Or as Charles Galton Darwin, the grandson of *the* Darwin put it, "A species is what a trained taxonomist says is a species."[386]

Appendix 3

Reproductive Isolation Barriers

One way to classify reproductive isolation barriers is to divide them into two general categories.

1. Those that prevent a male and a female from mating and producing a viable fertilized egg; these are called prezygotic barriers (a zygote is a fertilized egg).
2. Those that prevent a fertilized egg from maturing into an organism capable of further mating; these are called postzygotic barriers.

Prezygotic Barriers

1. Physical separation of habitats preventing mating—allopatric speciation.
2. Habitat isolation during breeding season. This is different from #1 in that the different species may occupy the same general habitat and location normally but develop some type of physical separation at the time of breeding.
3. Temporal breeding isolation—the two species occupy the same habitat but breed at different times.
4. Anatomical isolation—the sexual organs of the male and female are incompatible for copulation or sexual arousal.
5. Gamete incompatibility—the sperm and egg are physiologically incompatible and do not form a zygote when in contact.
7. Behavioral isolation—the males and females are not attractive to each other because of appearance, behavior, or some other characteristic. This is sometimes called sexual preference isolation or sexual selection.

Postzygotic Barriers

1. Gestation inviability—the zygote either does not implant in the uterus properly, or aborts after implantation prior to fetus viability.
2. Offspring sterility—the zygote develops normally, but the resulting hybrid individual is sterile.
3. Postzygotic behavior or sexual inviability—the hybrid individual is less attractive to other potential mates owing to physical, behavioral, or other characteristics.

Appendix 4
Everything You Didn't Want to Know About Genetics

Chapter 5 on genetics only begins to scratch the surface of the complexity of our genome and epigenome. I tried to limit the summarization of these processes to those topics I believe are most relevant to the possible emergence of *Homo nouveau.* I have included this appendix solely to give the reader a taste of what is missing in chapter 5. It by no means completes the discussion.

A. Dominant and Recessive Inheritance

The concept of dominant and recessive genes was developed before we really understood how genes work. In retrospect, it is probably an unfortunate and misleading choice of terms. The dominant variant of a gene doesn't actually exert any control over the recessive variant, and in most cases it doesn't impact it in any way. Usually, *dominant* simply means the variant produces a protein that does something, and either the recessive variant doesn't produce the protein, or it produces a different protein that is less apparent in the presence of the protein produced by the dominant variant.

For example, in the case of human eye color, the part of the eye that gives it its color is called the iris. The normal tissue that makes up the iris reflects light in a manner that appears blue in color. There is no blue pigment; it is simply the way the iris tissue is structured that reflects blue light. Therefore, the default iris is blue. However, if a brown pigment protein is added to the iris, it will appear brown in color. The "big B" gene produces the brown pigment, and the "little b" gene does not. Having one set of genes from either parent that produces the brown pigment protein is sufficient to color the iris brown. That is why you can be either BB (homozygous) or Bb (heterozygous) and have brown eyes. The little b variant simply doesn't produce any brown pigment protein. In this case, big B dominates only in the sense that it produces a protein whereas the little b variant does not. (Actually, human eye color is more complicated than this, as explained below in appendix 4-B, but it was a familiar example to use to explain dominant and recessive inheritance.)

To further illustrate why the terms *dominant* and *recessive* are somewhat misleading, consider the gene for one of the hemoglobin proteins in humans. There is a variant of this gene that causes sickle cell anemia. The sickle cell variant of the gene still produces a protein that is mostly like normal hemoglobin but contains one different amino acid. This slight difference in the hemoglobin causes the red blood cells to become distorted (to sickle) in a way causing a severe disease, called sickle cell anemia. However, if you only have one copy of the sickle cell gene variant from one parent and a normal copy from the other parent (i.e., heterozygous), there is enough normal hemoglobin in the red blood cells to keep the red blood cells from sickling. So phenotypically, the heterozygous form is normal, and that person has sickle cell trait but not sickle cell disease. We therefore say that the sickle cell variant of the hemoglobin gene is recessive to the normal variant even though it is functioning the way a gene functions and is producing a protein.

But to further confuse the sickle cell story, it turns out that any red blood cells that contain the sickle form of hemoglobin are resistant to infection from malaria. That is true whether the person is heterogeneous or homogeneous for the sickle cell gene. That is, the sickle cell gene is dominant with respect to malaria resistance. So we have the same gene that can be either dominant or recessive depending on which phenotype we are talking about.

There are other examples that muddy the water regarding the dominant and recessive concept. For example, the human ABO blood type shows both dominance and something called codominance. The A allele codes for a one protein (A) that sits on the surface of red blood cells, and the B allele codes for a different protein (B) that does the same. The O allele doesn't code for any protein. Thus in either the AO or BO heterozygous state, both A and B are dominant because the red blood cells have the protein. Thus the blood type is either type A or type B. But if the person is heterozygous AB (i.e., received one of each allele from each parent), then neither is dominant. They are instead codominant because the red blood cells have both proteins, and the blood type is type AB.

There is another situation called incomplete dominance. This is common in flowers where a cross between a red flower and a white flower often produces a pink flower (i.e., the traits are blended). In humans, this occurs with a match between one parent with curly hair and one with straight hair. The child often has wavy hair—something in between. Apparently, the Lamarckian blending theory of inheritance was only incompletely debunked by Mendel.

In fact, there is much about Mendel's work that is not as clear-cut as Mendel made it appear. Subsequent authors in the early twentieth century even suggested that his results were too good to be true, suggesting that his data may have been manipulated to correspond to his theory.[387] Although the basic concept of Mendelian inheritance is correct, involving alleles for each gene inherited from each parent, the details are far more nuanced and complicated than Mendel's conclusions.

There is a new situation that *Homo sapiens* had created that Mendel could not possibly have conceived of or even understood. In chapter 12 on genetic engineering, I discuss the concept of gene drive. This totally changes the meaning of dominance in that gene drive does change its corresponding allele to be identical to itself. It really does dominate.

So far the discussion of dominant and recessive genes is related to the genes located on one of the twenty-two autosomal chromosomes. Thus, I have been talking about autosomal dominant and autosomal recessive traits. The sex chromosomes present a different situation. Females have two X chromosomes, and males have one X and one Y. The X chromosome is a fairly normal sized chromosome, having about two thousand genes. The Y chromosome is a runt, having only sixty to eighty genes mostly related to male sex anatomy and hormones. A small portion of the Y chromosome corresponds to the X chromosome and is called the pseudoautosomal portion of both chromosomes. It is only in that portion of the Y chromosome that there are the usual two copies of each gene for each allele. Most of the genes on the Y chromosome do not have a matching allele on the X chromosome. That is why the Y chromosome is used to trace the paternal lineage.

Obviously the Y chromosome is dominant over the X chromosome in the sense that if you have a Y, you are a male. If you don't, you are a female. In reality, there is one particular gene, called the SRY gene, that causes the embryo to become a male, so the SRY gene is dominant.

Unlike the twenty-two pairs of autosomal chromosomes, the sex chromosomes are unique in that the female contains a pair of Xs and no Ys, whereas the male has one of each. Therefore, genes on the X chromosome have the usual two genes for each allele in females, but there is only one gene for each X allele in the male (except for the small pseudoautosomal portion). In the female, one of each gene on the X chromosomes comes from each parent. In the male, all the genes on the X chromosome come from the mother. This creates a different inheritance pattern, which we call sex-linked. Thus we have sex-linked dominant and sex-linked recessive traits.

An example of a sex-linked recessive trait is one form of hemophilia called hemophilia A, or clotting factor VIII deficiency. The mutant gene fails to produce a protein necessary for normal blood clotting. In the female, if just one copy of the allele has the mutant gene and the other copy has the normal gene, there is enough of the normal protein produced for blood clotting, and this person is phenotypically normal. However, she is called a carrier for hemophilia A. One-half of her male children on average will inherit the mutant gene and have the disease, because males have only one X chromosome. One-half of her female children will be carriers of the disease. The only way that a female can have the disease is if both her father has the disease, and her mother is either a carrier or has the disease. This is rare. Figures A1 and A2 illustrate the sex-linked recessive pattern. Both figures are taken from hemophilia-information.com.[388] The first figure shows the mating of a normal father with a carrier mother. The second figure shows the mating of a hemophiliac father with a normal mother.

Figure A1

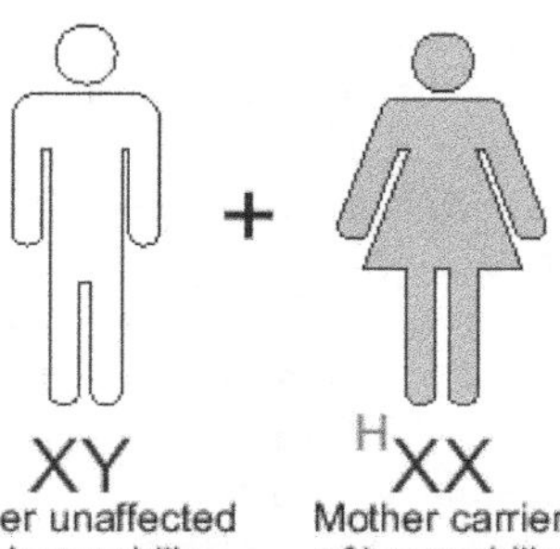

Possible outcomes for each pregnancy

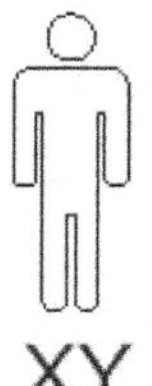

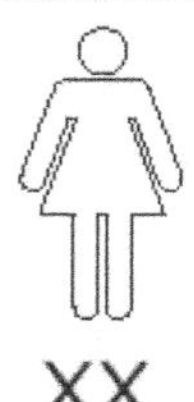

Figure A2

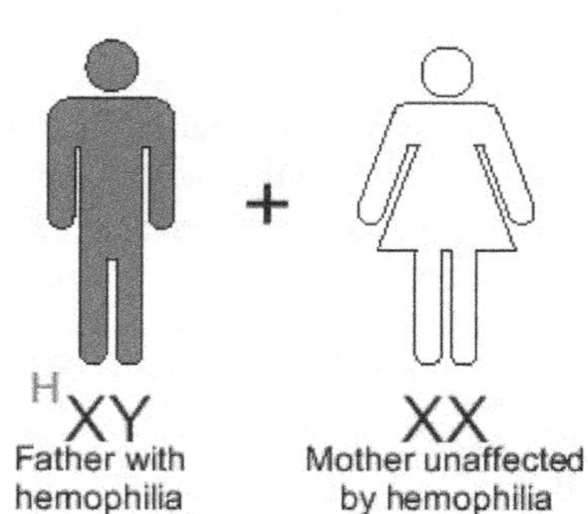

Possible outcomes for each pregnancy

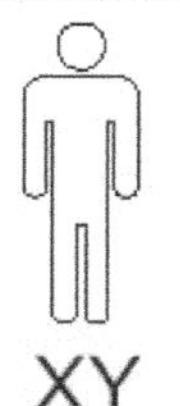

Figure A3 shows the inheritance pattern for a sex-linked dominant disease. If the father has the disease, all the daughters and none of the sons will inherit the disease. If the mother has the disease, half of the sons and half of the daughters will have the disease. There is no carrier state. Generally, sex-linked dominant diseases are rare, and the average reader will not likely have heard of any of them.

FIGURE A3

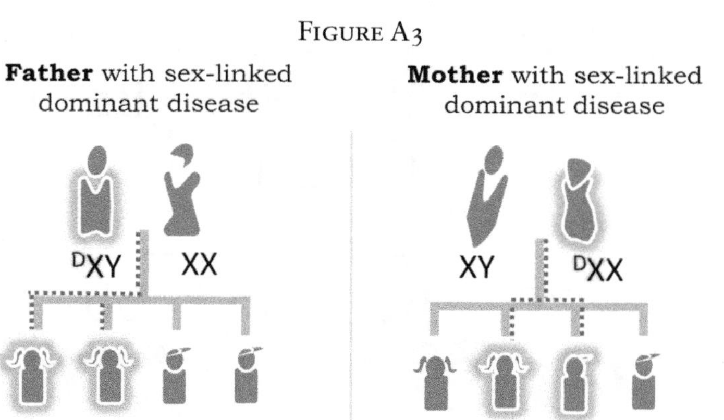

B. Human Eye Color Revisited

I've used the example of brown eyes and blue eyes as being a simple case of big B being dominant and little b being recessive. Figure A4 shows the classic matching of a homozygous big B with a homozygous little B. Although I've depicted the genotypes slightly differently, it is exactly the same situation as the green and yellow peas in figure 10.

FIGURE A4

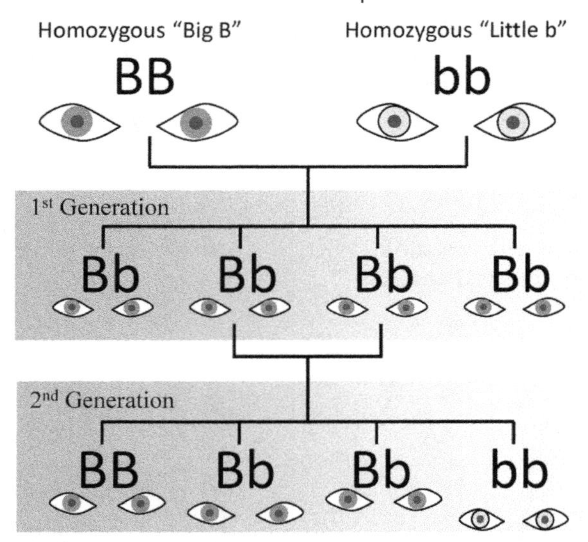

As usual, all of the first generation progeny are heterozygous and therefore are phenotypically brown-eyed. The mating of two heterozygous brown-eyed parents leads to the classic 75 percent brown-eyed and 25 percent blue-eyed distribution of progeny in the second generation. There is nothing new here. But I lied about the nature of the eye color gene. It really consists of two genes. Gene 1 and gene 2 are both required to produce the brown pigment. Each of these genes has a protein producing allele and a non-protein producing allele. Both proteins are required to produce the brown pigment. If there is a non-protein producing allele for either gene, no brown pigment is produced. This is shown in figure A5.

FIGURE A5

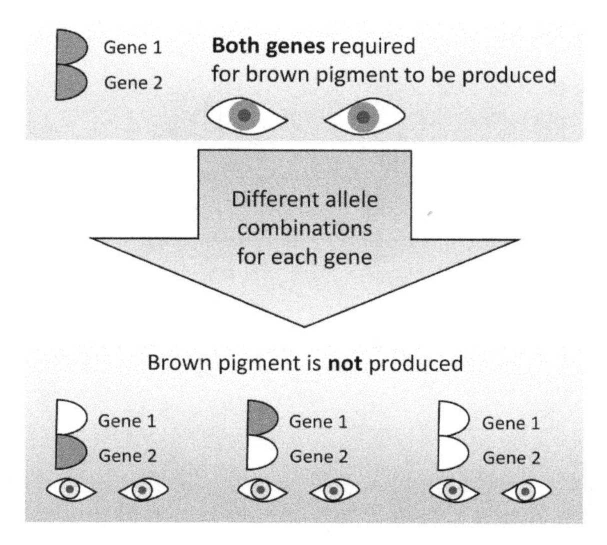

The reason I lumped the two genes together to make it appear as though they act as a single gene is because, in reality, that is basically true. These two genes are located very close together on their chromosome, and it is very unlikely that they would be separated during crossover recombination in meiosis.

In rare cases, it is possible that there could be a mating of two parents, one of which has the protein-producing allele in only gene 1 and the other parent having the protein-producing allele only in gene 2. Both parents would be phenotypically blue-eyed as shown in figure A6. However, there would be a 25 percent chance that they would have a brown-eyed child. That is because this child would be producing both required proteins for brown pigment, albeit from different genes of the chromosome pair.

Figure A6

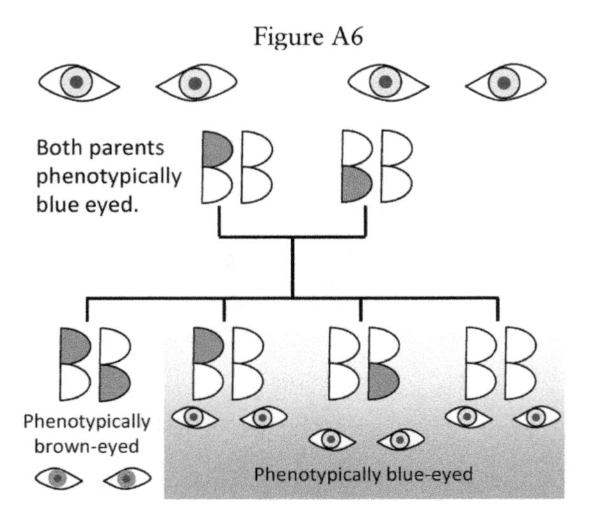

Although this is a rare situation, it does occur. To further complicate the human eye color situation, there is yet another gene that produces a pigment that makes the iris appear green. That's why we sometimes have people with green eyes or hazel eyes. Throw that into the mix, and the combinations and permutations become pretty large. Nothing is simple in genetics.

This phenomenon of two separate genes appearing to be inherited as one is related to the fact that they are physically adjacent or very close on the same chromosome, and thus they're unlikely to be separated during crossover in meiosis. This fact was used in the past to determine whether or not two genes were on the same chromosome and, if so, how close they were to each other. If the two traits determined by the genes were always inherited as independent units, they were assumed to be either on two different chromosomes or very far apart on the same chromosome. When it was observed that the traits tended to be linked statistically in inheritance, the degree to which they were linked was used as an inverse measure of their distance apart on the same chromosome. Today, with direct genome sequencing, this indirect and less exact measurement of gene position is no longer necessary.

C. Alternative Splicing

A single gene can produce multiple proteins. To understand how this occurs, look at figure A7: (This figure is inspired by a similar figure in Nessa Carey's book *Junk DNA. A Journey through the Dark Matter of the Genome.*)

FIGURE A7

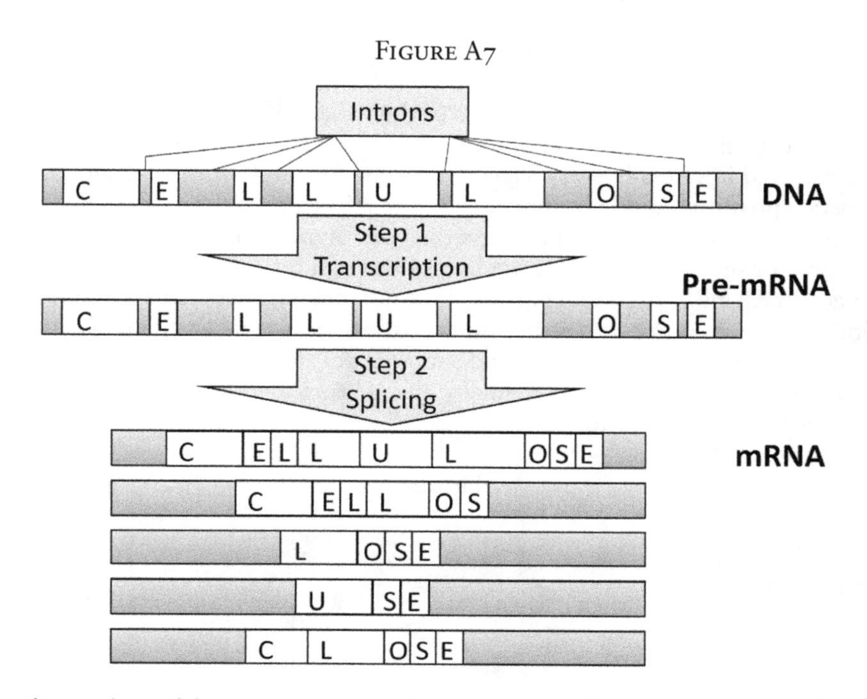

The top line of figure A7 represents a gene for the protein called CELLULOSE. The boxes around each of the letters represent the coding portion of the gene, and the line between the boxes represents junk DNA. The first remarkable thing about this is that a coding gene is not a continuous segment of DNA. The coding portions are broken up along the gene into smaller segments of various sizes, separated by junk. The portions of DNA between the coding segments are called introns. In step 1, the DNA for the gene is transcribed in its entirety into RNA (called pre-mRNA), including all of the introns. In the second step, the introns are eliminated producing an mRNA (messenger RNA) for the protein-coding portion of the gene for CELLULOSE. This is called splicing. The group of mRNAs and proteins involved in splicing is called the spliceosome.[389] Note also that there are a variety of other mRNAs produced by this same gene corresponding to other proteins: CELLOS, LOSE, USE, and CLOSE. This is the alternative splicing.

This raises a semantic question regarding the definition of the word gene. The above discussion is about a gene for the protein CELLULOSE. What do we call the portion of the gene that produces the protein CELLOS? Is it also a gene, or is it a subgene?

Introns sometimes play a role in gene regulation and other epigenomic functions, so they are not always junk. We also know that some diseases are caused by mutations in the introns.

It turns out that there is another genetic mechanism—other than alternative splicing—to produce multiple proteins from the same DNA sequence. It is called RNA editing and is used heavily in the octopus.[390] Since it is complicated and not relevant to humans, I will not discuss it further.

D. Epigenomic RNAs

The numbers, names, and functions of the variety of RNAs that do not code for proteins increase with every publication cycle. These non-coding RNAs are involved in virtually every aspect of genetic function. Without them, our genetic apparatus could not function. Figure A8 is a partial list of them. They are arbitrarily divided into small (less than two hundred bases) and long (two hundred or more bases). There are tens of thousands of them in the human epigenome. Unlike the coding genes, these non-coding genes have much fewer similar counterparts in other species, are expressed in smaller numbers, and are more difficult to detect. Our understanding of the role of RNA in health and disease is expanding exponentially and is beginning to enter into our therapeutic armamentarium.[391]

FIGURE A8

RNA type	Full Name	Function
Small RNA		
miRNA	Micro RNA	Silences mRNA
rRNA	Ribosomal RNA	Protein synthesis
scRNA	Small Cytoplasmic RNA	Protein transport
snRNA	Small Nuclear RNA	Alternative Splicing
snoRNA	Small Nucleolar RNA	rRNA biogenesis
tRNA	Transfer RNA	Amino acid transfer for protein synthesis
siRNA	Short Interfering RNA	mRNA expression
RNAi	RNA interference	mRNA expression
piRNA	Piwi Interacting RNA	Gene silencing in germ cells
Long RNA		
lncRNA	Long non-coding RNA	Various, gene regulation, slicing
asRNA	Antisense RNA	Gene regulation
lincRNA	Long Intergenic Non-coding RNA	Protein synthesis, stem cell regulation
ncRNA	Non-coding RNA	Miscellaneous category, multiple functions

E. Codon Table

FIGURE A9

1st base	2nd base								3rd base
	T		C		A		G		
T	TTT	(Phe/F) Phenylalanine	TCT	(Ser/S) Serine	TAT	(Tyr/Y) Tyrosine	TGT	(Cys/C) Cysteine	T
	TTC		TCC		TAC		TGC		C
	TTA	(Leu/L) Leucine	TCA		TAA	Stop (*Ochre*)	TGA	Stop (*Opal*)	A
	TTG		TCG		TAG	Stop (*Amber*)	TGG	(Trp/W) Tryptophan	G
C	CTT		CCT	(Pro/P) Proline	CAT	(His/H) Histidine	CGT	(Arg/R) Arginine	T
	CTC		CCC		CAC		CGC		C
	CTA		CCA		CAA	(Gln/Q) Glutamine	CGA		A
	CTG		CCG		CAG		CGG		G
A	ATT	(Ile/I) Isoleucine	ACT	(Thr/T) Threonine	AAT	(Asn/N) Asparagine	AGT	(Ser/S) Serine	T
	ATC		ACC		AAC		AGC		C
	ATA		ACA		AAA	(Lys/K) Lysine	AGA	(Arg/R) Arginine	A
	ATG[A]	(Met/M) Methionine	ACG		AAG		AGG		G
G	GTT	(Val/V) Valine	GCT	(Ala/A) Alanine	GAT	(Asp/D) Aspartic acid	GGT	(Gly/G) Glycine	T
	GTC		GCC		GAC		GGC		C
	GTA		GCA		GAA	(Glu/E) Glutamic acid	GGA		A
	GTG		GCG		GAG		GGG		G

F. Three-dimensional Importance of DNA and Proteins

The DNA in our chromosomes consists of two strands bound together in a double helix. Figures A10 and A11 are two typical images depicting this structure

FIGURE A10

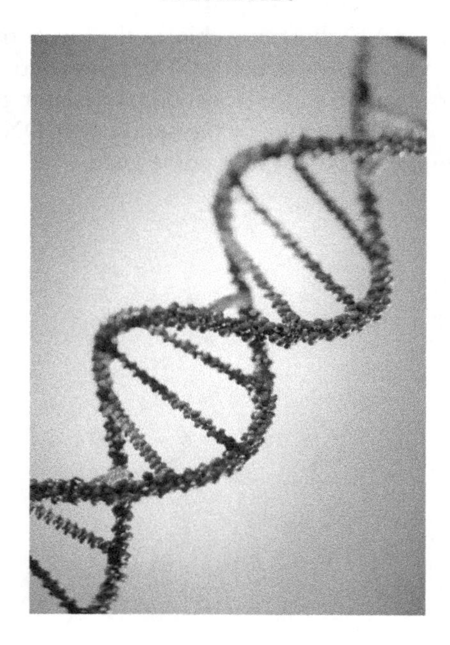

FIGURE A11

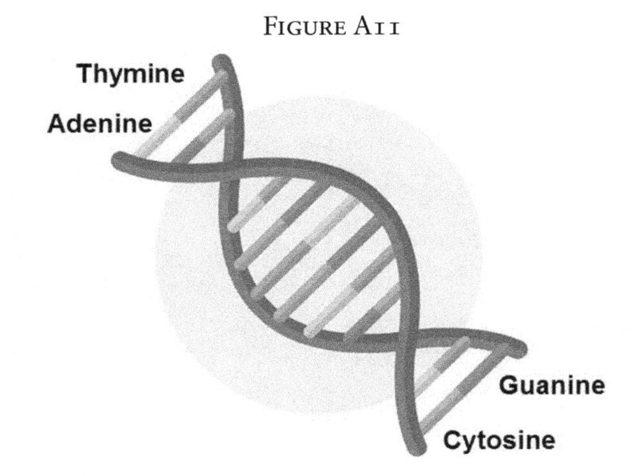

What these images don't show is the true shape of DNA. If our chromosomes were strung out like a string, they would be about two meters long—hardly able to fit in the nucleus of a single cell. Instead, they are wound up tightly around proteins called histones and folded repeatedly upon themselves into a complex, three-dimensional shape. Probably it would look more like a microscopic, tightly wound bowl of spaghetti. This allows the chromosomes to fit inside our nuclei, and the shape plays an important role in genetic functions. The folding brings different parts of the genome into juxtaposition with each other that otherwise would be far apart in the DNA sequence. This three-dimensional proximity is important in gene regulation. (A nice video of this can be seen at http://learn.genetics.utah.edu/content/epigenetics/intro.) Figure A12 illustrates this. In the top of the figure, the chromosome is shown as a linear structure. The protein-coding gene A is one million base pairs away from the noncoding regulator gene for gene A. However, as shown in the bottom of the figure, after DNA folding, the regulator gene is in juxtaposition to the coding gene.

FIGURE A12

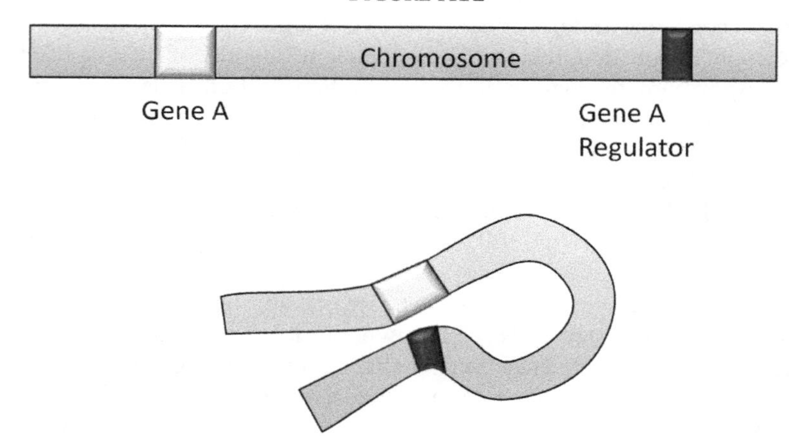

Likewise, the three-dimensional configuration of proteins is critical to their function. The shape determines their ability to fit with other molecules for the binding required to interact properly. Our epigenome has an elaborate set of RNA functions to control the exact nature of protein folding as the protein is being constructed in our ribosomes. In fact, the same protein controlled by a single gene can result in different folding patterns for different purposes depending on which codon version is being used for the amino acids.[392]

There is another interesting variant to DNA shape in humans. In the chapter on genetic engineering, circular segments of DNA called plasmids in bacteria were important in the early study of genetic engineering. We are now learning that there are circular segments of DNA in some human cells.[393] These segments are located in the nucleus of the cells and are called extra-chromosomal circular DNA or eccDNA. They are most commonly found in cancer cells but also rarely in normal cells. In cancer cells, they contain extra copies of cancer-promoting genes and may play an important role in cancer development. At this time, there is a lot of speculation on the role of human eccDNA in both health and disease and this new field of study is referred to as the "circulome."

G. The Barr Body

Females contain two X chromosomes, and males have one. The X chromosome has about two thousand coding genes. In the male, the one copy of these protein-producing genes produces exactly the right amount of protein for normal functioning. Having two copies of the same protein-producing gene is often deleterious, and this is usually moderated when that occurs in the other twenty-two pairs of autosomal chromosomes by epigenomic processes. However, in the female sex chromosomes, this potential problem is taken care of en masse by entirely deactivating one of the two X chromosomes in each cell. It does this through an epigenomic RNA-producing gene called Xist. This is very efficient. The Xist RNA (a long non-coding RNA; see appendix 4-D) essentially coats one and only one of the two X chromosomes, causing it to shrink down into a tight, inactive ball attached to the side of the nucleus of the cell. Approximately half of all the cells in females bodies have an active X from their mother, and the other half have an active X from their father. The shrunken inactive X structure can be seen under a microscope in all female cells and is called the Barr body. This has been used in the past to verify the female sex of athletes in the Olympics and other venues by taking a swab of the cells from the buccal (mouth) mucosa lining and looking at them under a microscope.

Exactly how Xist knows how to inactivate one and only one of the two X chromosomes in female cells is unknown, but it is very good at its job. Occasionally, there are males who have three sex chromosomes in the XXY configuration, called Klinefelter's Syndrome. Xist inactivates one of the two X chromosomes even in these males, so these are males who have Barr bodies. Sometimes females have only one X chromosome (Turner's Syndrome), X0, and Xist leaves that sole X chromosome alone. This is a female without Barr bodies. Other females have XXX, and Xist knows to inactivate two of them, leaving two Barr bodies in each cell.

Xist also explains why calico cats are almost always female. They are heterozygous for genes that cause an orange pigmentation in fur and a black pigmentation in fur. These genes are on the X chromosome, and because 50 percent of the X chromosomes are inactivated randomly in all cells, a cat that inherits both the orange and black alleles has a beautiful mosaic fur pattern.

FIGURE A13: CALICO CAT

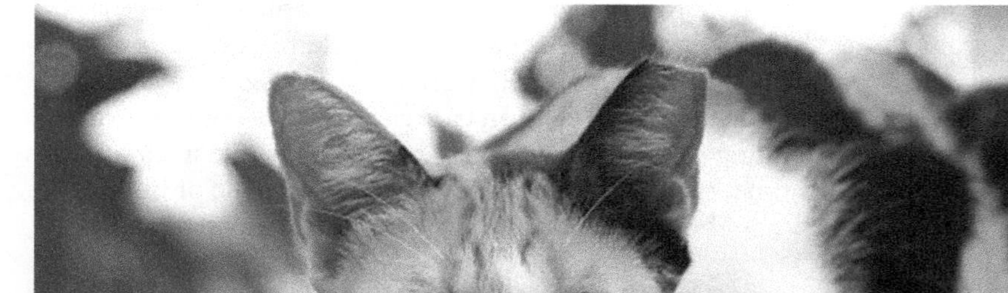

Why is there any white fur on this cat? There is another gene on an autosomal chromosome called the Piebald gene, which causes white fur. It is dominant over both the orange and black X-linked alleles. However, its expression or penetrance is very variable from cell to cell, and thus only part of the cat is white. Don't ask me how that works.

H. Interesting Facts about Mitochondria

1. All eukaryotes, which include all plants and animals, have mitochondria in each cell.
2. Mitochondria evolved billions of years ago by early eukaryotes engulfing prokaryotic organisms (like bacteria) and retaining them within each cell. These prokaryotes then became the mitochondria.
3. There are scores of mitochondria in each cell. They replicate many times between each cell division.
4. Mitochondria in humans contain thirty-seven genes, thirteen of which are protein-coding genes for enzymes involved in the energy producing functions of the cell. The remaining genes code for various types of RNA also related to energy functions. This compares to over twenty thousand nuclear genes.
5. Originally, mitochondria contained many more genes than the current thirty-seven, but most of these migrated to become incorporated into nuclear DNA. Thus, many nuclear genes produce proteins and RNA that relate to mitochondrial function.
6. Mitochondria are only passed onto progeny from the egg of the mother. Mitochondria in sperm cells are lost during fertilization of an egg. Initially, it was thought that the sperm mitochondria were sloughed with the tail, however it is now believed they actually are destroyed after fertilization.
7. Because there are many mitochondria per cell, and because they replicate many times before each cell division, there is a higher volume of mitochondrial gene copies in each cell than nuclear genes; nuclear DNA only replicates once with each cell division. That is why it is easier to find mitochondrial DNA in fossils than nuclear DNA.
8. Mitochondrial DNA is double-stranded like nuclear DNA, but unlike nuclear DNA, it is circular.
9. Because of the higher rate of replication of mitochondrial DNA than nuclear DNA, there is a higher rate of mutation in mitochondrial DNA.
10. Some researchers think that the faster mutation rate of mitochondrial genes compared to nuclear genes leads to incompatibilities with the related nuclear genes; this induces faster mutation rates in those nuclear genes.
11. There is evidence to suggest that the vivid red color of some birds correlates with compatibility between the mitochondrial and nuclear genes, and therefore it leads to sexual selection of birds with healthier mitochondria.
12. There is speculation that the emergence of meiosis occurred because of the incompatibilities in rates of evolution between mitochondria and nuclear

DNA. The crossover recombination associated with meiosis increases genetic mixing and nuclear evolution.
13. There are certain diseases caused by mutations in mitochondrial genes. See appendix 5 for a discussion of mitochondrial gene therapy.
14. Mitochondrial genetic diseases affect approximately 1/5000 people.
15. The genetic makeup of mitochondria in any given cell varies from mitochondrion to mitochondrion. Only a small proportion of the maternal mitochondria are passed on to any given female germ cell, thus it is impossible to predict the exact genetic make-up of any given egg from the maternal mitochondrial DNA mix. Therefore, a mother with mitochondrial disease will not necessarily pass on the genetic disorder to her offspring.

I. Adam and Eve

Every female *Homo sapiens* alive today has mitochondrial DNA that can be traced back to a single female that lived in Africa some time between 100,000 and 200,000 years ago. This single female is often referred to as "Mitochondrial Eve." The reason for this is that mitochondria are passed only to progeny from the mother; thus Mitochondrial Eve is tracing only the matrilineal ancestry—not the patrilineal ancestry. This does not imply that Mitochondrial Eve was the only female that existed at that time (unlike the biblical Eve). There were other female *Homo sapiens* alive at that time, however, their female lineage died out at some later time either because one of the female descendants had no children or only male children. Mitochondrial Eve's mitochondrial DNA represent the only remaining continuous female lineage. Further, there are male descendants alive today of females other than Mitochondrial Eve; however, the lineage is through at least some males thus breaking the mitochondrial DNA chain. Finally, the mitochondrial DNA of females today is not identical to the DNA of Mitochondrial Eve because mutations have occurred from time to time. In fact, the variations in the DNA related to these mutations is used to categorize all living females into various groups which allows their ancestry to be related to various geographical regions as well as into time periods since the various mutations occurred. The rate of mutations is used to estimate the time of Mitochondrial Eve's existence. There is a range of error associated with mutation rate calculations, which accounts for the wide range associated with that time.

An analogous calculation can be made for all living males today based on the nuclear Y sex chromosome, which is only passed on from the father. All of today's living males can be traced back through the Y chromosome to an ancestral "Adam" who also lived in Africa. Again the time of the existence of Y chromosomal Adam is uncertain; it was likely earlier than Mitochondrial Eve by as much as 140,000 years. As with the mitochondrial DNA, various mutations in the Y chromosome allow classification of today's males in various groupings.

Appendix 5

Mitochondrial Gene Therapy in Humans

There is a class of diseases that are caused by mutations in mitochondrial genes.[394 395] One set of these mutations is associated with an increase in spontaneous abortions following *in vitro* fertilization. In the 1990s, Dr. Jacques Cohen at the St. Barnabus Institute in New Jersey pioneered a technique in which a donor egg was used to replace the mitochondria of a mother undergoing *in vitro* fertilization.[396] This technique is illustrated in figure A14.

FIGURE A14

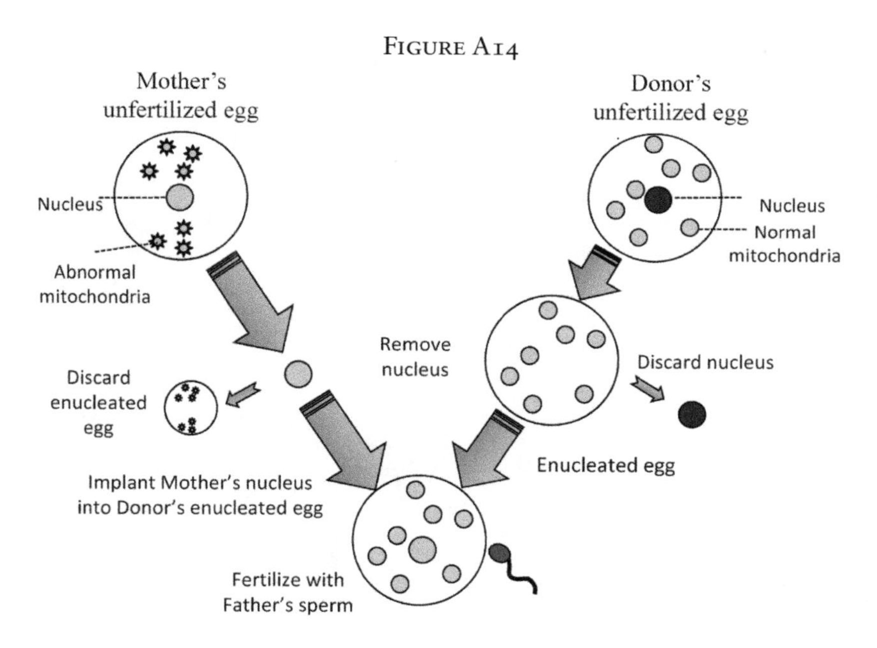

The resulting child has its entire nuclear DNA from its biological parents and only the mitochondrial DNA from a third-party donor. In essence, the child has three genetic parents. This technique modifies the DNA in a germline cell, an egg, that subsequently is fertilized and implanted in a womb, and so the mitochondrial DNA of the donor is passed on not only to the child but also to the child's progeny if the child is a female.

It is estimated that between thirty and fifty such procedures were done worldwide in the late 1990s and early 2000s. One of the children developed a cognitive disorder similar to autism. Although it was not known whether that was related to the mitochondrial procedure, because of the uncertainty and extreme controversy over any type of germline cell therapy, in 2002 the FDA decided to require all future such procedures undergo an investigational new drug application. That, essentially, put a halt to the procedure in the United States. However, in March 2015, mitochondrial gene therapy became legal in Great Britain.

Appendix 6

Fossil Hominin Species

1. *Sehelanthropus tchadensis*, also called Toumai. Discovered in Chad. Seven million years old. Candidate for earliest hominid.

2. *Orrorin tugenensis,* also called Millennium Man. Discovered in Kenya. Six million years old. Candidate for earliest hominid.

3. *Ardipithecus kadabba*, discovered in Ethiopia. 5.8 million years old. Candidate for earliest hominid.

4. *Ardipithecus ramidus,* also called Ardi and *Australopithecus ramidus.* Discovered in Ethiopia and Kenya. 4.4 million years old. Discovered by Tim White. Currently most widely accepted earliest species in Hominidae lineage. Bipedal on ground with opposable big toe for tree climbing.

5. *Australopithecus anamensis,* discovered in Kenya and Ethiopia. 4.1 million years old. Ancestor to Lucy.

6. *Australopithecus afarensis,* discovered in Ethiopia and Tanzania. 3.7 million years old. Most famous example is the fossil, "Lucy", discovered by Donald Johanson. Almost fully bipedal. No opposable big toe.

7. *Australopithecus bahrelghazali,* also called Abel. Discovered in Chad. 3.6 million years old. May be same species as Lucy. Only *Australopithecus* in Central Africa.

8. *Kenyanthropus platyops,* discovered in Kenya. 3.3 million years old. Could be classified as *Australopithecus.* May even be same species as Lucy. Could be first tool user.

9. *Australopithecus africanus,* also called Taung Child. Discovered in South Africa. 2.8 million years old. First discovered in 1924 by Raymond Dart, therefore first *Australopithecus* fossil discovered.

10. *Paranthropus aethiopicus,* also called *Australopithecus aethiopicus, Paraustralopithecus aethiopicus* and OMO 18. Discovered in Ethiopia. 2.6 million years old. With others in Paranthropus genus, called robust australopithecines because of a more "robust" jaw indicating more fibrous diet. Earlier australopithecines are called "gracile" australopithecines.

11. *Australopithecus garhi,* discovered in Ethiopia. 2.5 million years old. Another Tim White discovery. Could have used tools.

12. *Paranthropus boisei,* also called *Australopithecus boisei, Zinjanthropus boisei,* Zinj and Nutcracker Man. Discovered in Tanzania. 2.3 million years old. Also alleged to be first tool user.

13. *Paranthropus robustus*, also called *Australopithecus robustus.*

Discovered in South Africa. Two million years old.

14. *Australopithecus sediba.* Discovered in South Africa. Two million years old. Most advanced *Australopithecus.*

15. *Kenyanthropus rudolfensis*, also called *Pithecanthropus rudolfensis, Homo rudolfensis* and *Australopithecus rudolfensis.* Discovered in Kenya. 2.1 million years old. Lots of controversy regarding its classification. Overlapped with *Homo habilis.*

16. *Homo habilis,* also called *Australopithecus habilis.* Discovered in Tanzania. 2.2 million years old. Arguably the first *Homo* species. Overlapped with other *Homo* species and australopithicines. Used early tools for carving game. Brain size intermediate between australopithicines and *Homo egaster.*

17. *Homo egaster,* discovered in Eastern and Southern Africa. 1.8 million years old. Either ancestor to or the same as *Homo erectus.* In lineage to later *Homo* species including *Homo sapiens.*

18. *Homo georgicus,* discovered in Georgia. 1.8 million years old. Seems to be intermediate between *Homo habilis* and *Homo erectus.*

19. *Homo antecessor,* also called *Homo mauritanicus.* Discovered in Europe. 1.2 million years old. Possibly intermediate between *Homo egaster* and *Homo heidelbergensis.* Some consider *Homo antecessor* to be the European version of *Homo egaster* and *Homo mauritanicus* to be the African version of *Homo egaster.*

20. *Homo erectus,* discovered in Africa and Asia. 1.9 million years old. One of the longest surviving extinct *Homo* species in the *Homo sapiens* lineage. Led the first "Out of Africa" wave into the Middle East, Europe and Asia leading to *Homo heidelbergensis* and then *Homo neanderthalensis* and *Homo denisova.* In Africa, became *Homo heidelbergensis* and then *Homo sapiens.*

21. *Homo heidelbergensis,* discovered in Africa, Europe and Asia. 600,000 years old. Direct precursor to *Homo neanderthalensis, Homo denisova* and *Homo sapiens.*

22. *Homo helmei,* discovered in South Africa. 250,000 years old. May be intermediate between *Homo heidelbergensis* and *Homo sapiens* or a variant of either one.

23. *Homo floresiensis*, also called The Hobbit. 100,000 years old. Unusual recent *Homo* fossil found on the Island of Flores in Indonesia. Was only 3.5 ft. tall. Some consider it an unusual variant of *Homo sapiens* or perhaps a diseased form. Considered by some a subspecies of *Homo sapiens.* Survived until 12,000 years ago.

24. *Homo neanderthalensis,* discovered in Europe and Asia. 500,000 years old. Very close relative of *Homo sapiens* but not in the direct lineage. Larger bodies and brains than *Homo sapiens*, culturally very advanced but lacking the more advanced cognitive capabilities of Cro-Magnon *Homo sapiens* with whom they overlapped until about 30,000 to 40,000 years ago in Europe

and with whom they interbred to a small extent. They also overlapped with *Homo denisova* in Southern Asia.

25. *Homo denisova,* discovered in Asia. 400,000 years old. Known only by the DNA from a small finger bone fossil found in a cave in Denisova in Southern China. Very closely related to both *Homo neanderthalensis* and *Homo sapiens* and interbred with both to a small extent. Not in direct lineage to *Homo sapiens.*

26. *Homo sapiens idaltu*, discovered in Ethiopia. 160,000 years old. Considered to be the only subspecies of *Homo sapiens* other than *Homo sapiens sapiens* (with the possible exception that some might call *Homo floresiensis* a subspecies of *Homo sapiens.)*

Appendix 7
Tools to Study the Brain

1. **Tools to Study Brain Structure**

A. CT—Coaxial Tomography (also called Computerized Axial Tomography (CAT): Utilizes x-rays and computer processing to create cross-sectional images (slices) of the brain. These image slices can be superimposed in a computer to create three-dimensional images of the brain.
Advantages: non-invasive, more detail than simple x-ray.
Disadvantage: exposure to radiation.

B. MRI—Magnetic Resonance Imaging: Utilizes powerful magnets to elicit radio-frequency waves from human tissues, which are used to create images of the brain.
Advantages: noninvasive, no exposure to radiation, safe, generally higher resolution images than CT.
Disadvantages: expensive, claustrophobic, cannot be used in patients with metal implants or cardiac devices.

C. Post-mortem dissection: This is the original manner in which brain structure was studied before modern imaging techniques. It is still used to provide supplemental information at the time of autopsy.
Advantages: Very accurate, more detail, allows special staining
Disadvantages: Costly, time consuming, does not reflect live activity

2. Tools to study brain function.

A. EEG—Electroencephalography: Detects electrical activity of the brain using sensors attached to the scalp. Useful in studying epilepsy, sleep disorders, certain other brain conditions, and diagnosing brain death.
Advantages: non-invasive.
Disadvantages: low positional and structure resolution.

B. PET—Positron Emission Tomography: This is an imaging tool that detects certain metabolic activities in the brain. A radioactive substance is injected into the blood stream; this substance in taken up by cells with high metabolic activity related to that substance. The higher concentration of the radioactivity in these cells shows up as bright spots during PET-scan imaging. Depending on the substance used, this technique is useful in detecting cancers, certain dementias, certain brain diseases, as well as to study normal brain function.
Advantages: Measures brain activity rather than structure, non-invasive.
Disadvantages: Costly, some exposure to ionizing radiation.

C. fMRI—Functional Magnetic Resonance Imaging: This is a form of MRI that is able to detect blood flow in the brain, which reflects brain activity. The images therefore show which areas of the brain are active during various brain activities such as movements, doing math, and emotional reactions. It is a major technique for functional mapping of the brain.
Advantages: noninvasive, functional mapping.
Disadvantages: expensive, claustrophobic, cannot be used in patients with metal implants or cardiac devices.

D. MEG—Magnetoencephalography: A neuroimaging technique that measures naturally occurring magnetic fields produced by electrical currents in the brain. The images reflect brain activity and are useful in localizing brain areas during cognitive and other brain activities.
Advantages: non-invasive, high temporal and special resolution.
Disadvantages: needs special magnetic shielding, expensive.

E. Near-Infrared Spectroscopy: An imaging technique that measure the near-infrared spectrum of electromagnetic radiation. This spectrum reflects hemoglobin (blood) concentration in the brain and is useful in studying brain activity as reflected by blood flow. It has similar uses to fMRI.
Advantages: noninvasive, can be used on infants and moving subjects, portable.
Disadvantages: only measures blood flow in the cerebral cortex and not deeper brain tissues, as fMRI does.

F. Optogenetics: Specific brain cells are genetically modified to contain special light-sensitive proteins that cause those cells to fire (activate) when exposed to a specific wavelength of light. Alternatively, other light-sensitive proteins can cause the neuron to be inhibited by light. Light is delivered to the cells through very thin fiberoptic cables.
Advantages: the specific functions and activities of specific brain cells can be studied.
Disadvantages: requires invasive placement of fiberoptic cables. Complex technology to perform

G. Sonogenetics: Similar to optogenetics, except that the genetic modification of the brain cells make them sensitive to specific wavelengths of sound rather than light. This eliminates the need for the invasive placement of the fiberoptic cables used in optogenetics.

H. ExFISH: Expansion Fluorescent in situ Hybridization imaging of RNA—Allows nanoscale precision identity and visualization of RNA molecule locations within brain (and other) tissues in cultured and pathology brain samples. This is useful in studying normal RNA functions as well as certain neurological disorders affecting RNA.

Appendix 8

Explanation of Gene Comparisons

In chapter 6, I explained how we have determined that the chimpanzee is our closest living relative by comparing our genes. Figure A15 illustrates this. Let me assume a hypothetical species that has just five genes, labeled A through E (Generation 1). Assume that gene E mutates to F and splits off as a new species (Generation 2). Assume that each of the species at Generation 2 splits off another new species at Generation 3, with one additional genetic mutation in a different gene. Finally, do the same thing to create Generation 4. Are you still with me? Now compare Species 2 at Generation 4 with Species 3 closest to it in evolution. They overlap in four of the five genes. But now compare Species 2 to Species 4 further away in the evolutionary tree at Generation 4. None of the genes overlap in this small example. Note also that Species 2 differs from the original Species 1 in its lineage by only two genes (i.e., three of the genes overlap). Thus, in just three generations involving only five genes, one can see the power of doing the gene comparison. There is quantitatively more overlap with species closest to them in the evolutionary tree, but they also retain overlap with their ancestors. Now multiply this by thousands of genes and billions of nucleotides, with each gene having a variable number of mutations, and you can understand why computer modeling is now used to sort this out. Complicating this further, not every individual of a single species has the same mutations. That is why it is essential to do these studies on as many individuals in a population as possible, in order to obtain statistically valid sample sizes. This is easier to do in living species, and it's almost impossible in extinct species.

Figure A15
Gene Comparisons

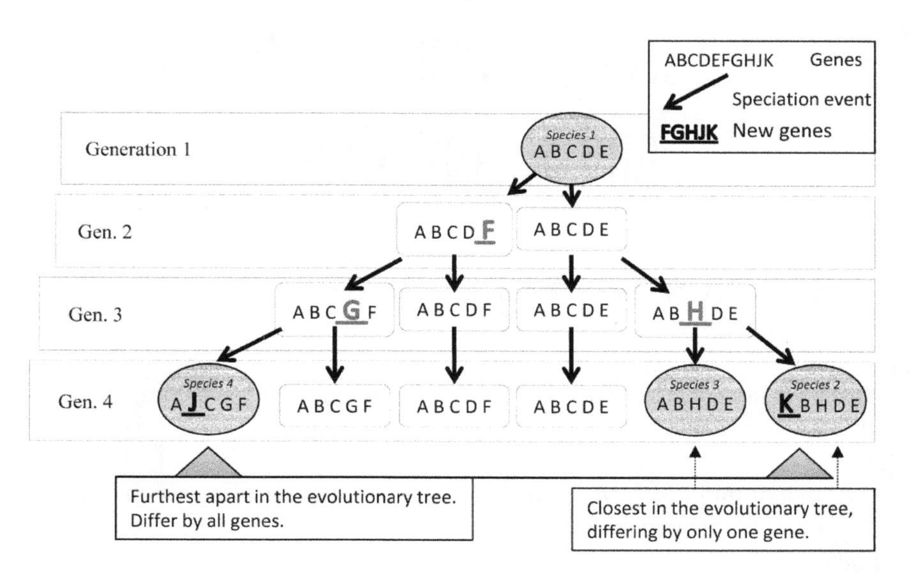

Glossary of technical terms and abbreviations.

AGI
Artificial General Intelligence—the simulation in a computer of generally all human intelligence.

AI
Artificial Intelligence—the simulation in a computer of some component of human intelligence.

Allele
Any one of multiple forms of a single gene, e.g. the gene for pea color has two types of alleles, one for yellow pea color and one for green pea color.

Allopatric
Originating in or occupying different geographical or physical areas. Contrasted with sympatric.

Alternative splicing
The process whereby a single gene can produce multiple proteins.

Amino acids
Chemical compounds that are the building blocks of proteins. Of the over 500 types of amino acids in nature, only 20 are used to build the proteins in living organisms.

Anagenesis
Evolution of a species without branching of the evolutionary line of descent. Compare to cladogenesis.

Anthropology
The study of the origins, physical and cultural development, biological characteristics, and social customs and beliefs of humans.

Archaea
Single-celled organisms that constitute one of the three major domains of living organisms. The other two domains are Bacteria and Eukarya. Archaea differ from bacteria primarily in their cell wall structures.

Archaic *Homo sapiens*
The group of species in the *Homo* genus that preceded *Homo sapiens* in the evolutionary lineage.

Ardipithecus
A genus of extinct Hominidae that lived 4-6 million years ago. They may be in the *Homo* lineage, walked upright when on the ground but had an opposing big toe for better tree navigation.

Artificial Intelligence
The simulation in a computer of some component of human intelligence.

Artificial superintelligence
The simulation in a computer of intelligence that exceeds human intelligence.

ASI
Artificial superintelligence—the simulation in a computer of intelligence that exceeds human intelligence.

Australopithecus
A genus of extinct Hominidae that lived 3-4 million years ago. Lucy is the most famous fossil. Walked upright without an opposing big toe. May be the precursor to the *Homo* genus.

Autosomal
Pertaining to the chromosomes that are not sex-linked. In humans, there are 22 pairs of autosomal chromosomes and one pair of sex-linked chromosomes.

BBI
Brain-to-Brain Interface—a group of technologies that allow humans and other animals to communicate using direct brain sensors and brain stimulators.

Bipedal
Terrestrial locomotion using two legs.

CAR-T
Chimeric Antigen Receptor T-cells—these are genetically engineered immune cells created to treat certain cancers.

Cas9
CRISPR Associated Protein 9—an enzyme used in conjunction with CRISPR in genetic engineering to cut DNA. Other enzymes, such as Cpf1, have also been discovered which serve the same purpose as Cas9.

Chromosomes
The threadlike structures within the nucleus of eukaryote cells containing the genetic DNA. Humans have 23 pairs of chromosomes in each cell.

Cisgenic
Genetic modification containing only genes from within the same species.

Cladogenesis
An evolutionary splitting event when a parent species splits into two distinct species.

Coding gene
A gene that defines the template for a protein or set of proteins. In common usage, a gene is the same as a coding gene.

Codon
A set of three nucleotides defining the template for a specific amino acid or the "stop" instruction in protein synthesis directed by DNA. There are 64 codons in the genome of organisms.

Connectome
A comprehensive map of neural connections in the brain and other components of the nervous system of an organism.

CRISPR
Clustered Regularly Interspaced Short Palindromic Repeats—This is a DNA sequence found naturally in bacteria which, in conjunction with Cas9, serves as an immune defense mechanism again viruses. Modifying the CRISPR sequence has become a powerful tool for genetic engineering.

Crossover recombination
The creation of new combinations of alleles during meiosis formed when the copying mechanism switches randomly between the corresponding chromosomes of each parent.

Darwinian evolution
The process of species modification during evolution as described by Charles Darwin in which environmental factors determine which combinations of genes provide traits (phenotype) that best lead to survival and reproductive success. This process is called natural selection.

Denisovan
A member of an extinct species of humans living in Asia at the same time as the Neanderthals known only by genetic analysis of a few small fossil fragments found in the Denisova Cave in Siberia.

Deoxyribonucleic acid
A string of permutations of four nucleotides that constitute the genetic material. The nucleotides are cytosine, thymine, guanine and adenine. Each sequence of three nucleotides constitutes a codon defining the inclusion of a specific amino acid into a protein.

DNA
Deoxyribonucleic acid

Domain
The highest level of taxonomy consisting of three domains: Archaea, Bacteria and Eukarya. All living organisms fall into one of these three domains. Domains are further divided into Kingdoms, Phyla, Classes, Orders, Families, Genera and Species.

ECoG
Electrocortography—a technology in which electrodes on the surface of the brain record signals that are communicated to a computer and interpreted to perform various functions.

EEG

Electroencephalography—a technology in which electrodes on the surface of the skull record brain wave signals. This non-invasive technology has been used for decades to diagnose epilepsy, sleep disorders and other brain conditions.

Epigenetics

The study of the epigenome.

Epigenome

The set of chemical processes that affect the expression of genes. These consist of various noncoding genes and chemical mechanisms including various RNA sequences, methylation, and others. The epigenome does not alter the DNA sequence.

Eukarya

One of the three domains of life characterized by species with cells containing a nucleus surrounded by a nuclear membrane. All plants and animals are in this domain. The other two domains are Bacteria and Archaea, both consisting of single-celled organisms.

Eukaryote

A member of the Eukarya domain characterized by species with cells containing a nucleus surrounded by a nuclear membrane.

Evolution

Change in the heritable traits of biological populations over successive generations by such processes as mutation, natural selection, and genetic drift.

Evolutionary biology

The study of evolutionary processes.

Exaptation

The study of evolutionary processes. In evolution, a feature or trait of an organism that was originally adapted or selected for one function is later coopted to perform another function. For example, jawbones in fish were later adapted to be hearing ossicles in mammals.

Expression

As related to genes, the degree to which a gene is operational in any given cell at any given time in an organism's development. Gene expression is governed by the epigenome.

Family

The taxonomic level just above genus that usually includes multiple genera.

Fertilization

The union of a male and female gamete. In humans, this is the union of a sperm and egg to form a zygote.

Fictive thinking

The ability to communicate (fictive language) and think about (fictive thinking) fictitious or imaginary concepts and about concepts that have no objective reality such as countries, corporations and organizations. It is a defining characteristic of *Homo sapiens*.

fMRI

Functional Magnetic Resonance Imaging—a technology that allows visualization of blood flow in the brain that reflects brain activity. It is useful in mapping brain function as well as brain structure.

Fossil

The preserved remains or imprints of organisms from former geologic eras.

Gamete

A mature male or female reproductive cell that combines with a gamete of the opposite sex to form a zygote. A gamete has only one of each chromosome type (haploid) rather than a pair of each chromosome type (diploid). The combination of two gametes during fertilization produces a complete diploid DNA set.

Gene

The basic unit of heredity consisting of a DNA sequence. General usage implies a "coding" gene that defines the template for a protein or set of proteins. There are also "noncoding" genes.

Gene drive

A genetic engineering technique that leads to a particular gene allele having greater than the normal 50% chance of being inherited.

Gene pool

The set of all genes in the population of a particular species.

Genetic drift

The random change in the frequency of gene alleles in a population over time.

Genetic engineering

The direct manipulation of an organism's genome using biotechnology.

Genetics

The study of genes, genetic variation, and heredity in living organisms.

Genomics

The subset of genetics that focuses on the study of the structure and function of the genomes of organisms. DNA sequencing, data analytics, epigenomic mechanisms and studies of 3-D structure are all components of genomics.

Genotype

The set of DNA sequences that determine a cell or organism trait.

Genus

The taxonomic level just above species. A species is defined by the combination of its genus and species names.

Germline cells

Cells whose genetic makeup is passed on to the next generation. These include embryonic stem cells, gametes (egg and sperm in humans) and zygotes.

Heterozygous

Having two different alleles for the same gene.

Hominid

Any of the living or extinct bipedal primates of the family Hominidae, including all species of the genera *Homo, Australopithecus* and *Ardipithecus*.

Hominidae

A taxonomic family of genera that includes extant great apes and humans and multiple extinct genera including extinct *Homo* species, *Ardipithecus* and *Australopithecus*

Homo erectus

An extinct *Homo* species that lived from 1.8 million years ago to about 50,000 years ago—one of the longest surviving hominids both in Africa and out of Africa. Probably in the direct *Homo sapiens* lineage.

Homo heidelbergensis

An extinct *Homo* species that lived from 600,000 to 200,000 years ago and probably the direct ancestor to *Homo sapiens*, Neanderthals and Denisovans.

Homo nouveau

Imaginary future *Homo* species designated as "what comes after *Homo sapiens.*"

Homo sapiens

Today's modern humans. Emerged in Africa about 200,000 years ago and spread throughout the world.

Homozygous

Having both alleles being the same for a specific gene.

HSAM

Highly Superior Autobiographical Memory—a condition in which a person remembers every detail about their own lives but unlike Savant Syndrome individuals, does not have autism or other mental disorders.

Hybrid

The offspring of two different species.

Hybridization

The creation of a new species by the interbreeding of two species. This is rare in animals.

IA

Intelligence amplification—this is distinguished from AI (artificial intelligence) in that AI usually replaces intelligent functions whereas IA supplements or augments intelligent functions.

IVF
In vitro fertilization

Kingdom
In taxonomy, the classification just below the highest level (domain). Humans and all other animals are in the kingdom Animalia.

Knockdown gene therapy
In genetic engineering, the ability to disable, replace, or otherwise render inoperable a gene producing a deleterious effect.

Lamarckian evolution
The philosophy promulgated by Jean-Baptiste Lamarck, a French biologist, in which offspring manifest a blending of parental traits that were acquired during a lifetime. This theory was largely debunked and replaced by the concepts of Mendelian inheritance. We now know that some epigenetic changes to the genome as well as germline cell mutations, which are acquired during a lifetime, are passed on to offspring.

Lineage
A sequence of species each of which is considered to have evolved from its predecessor.

Malthusian
A person who believes that the rate of increase of a population should be controlled; otherwise the population will outgrow the resources needed to sustain it. Thomas R. Malthus, a British economist, promulgated this view.

Mass extinction
The extinction of more than 50% of living species in a geologically short period of time. There have been five such mass extinctions in the history of earth and some believe we are currently undergoing the sixth.

Meiosis
In sexually reproducing organisms, the process whereby specialized embryonic stem cells (in the testis and ovary of humans) go through special cell divisions to produce the gametes (sperm and eggs). In the process the fully paired complement of chromosomes (diploid) is reduced to half (haploid) in each gamete. Each haploid chromosome contains a mixture of alleles from each parent.

Mendelian genetics
The laws of inheritance as described by Gregor Mendel in which discreet inheritance units (later names genes) can be dominant or recessive and result in predictable ratios of offspring traits (later called phenotype) depending on the "purebred" or "hybrid" nature of the genes (later called the genotype). Traits are inherited as "all or none" rather than blended.

Messenger RNA
A large family of RNA molecules that convey genetic information from DNA to the ribosome, where they specify the amino acid sequence of a protein.

Metapopulation

A subset of the total population of organisms sharing a common gene pool.

Methuselarity

A term coined by Aubrey de Grey of the SENS Institute which indicates the point at which our ability to reduce the effects of aging exceeds the aging process. Essentially at that point, life expectancy increases indefinitely.

Methylation

When referring to DNA (DNA methylation), a process by which methyl groups are added to DNA. Methylation modifies the function of the DNA. When located in a gene regulator, DNA methylation typically acts to repress gene transcription and expression, usually permanently. Methylation is an epigenetic process that does not alter the DNA sequence.

Mitochondria

Small organelles found in the cytoplasm of most eukaryote cells. They provide the power or energy for cell metabolism. They contain a small number of genes that are inherited only from the mother.

Mitosis

In multicellular organisms, the process of somatic cells dividing into two identical new cells containing the full diploid complement of chromosomes. This is also how non-sexually reproducing single-celled organisms reproduce.

Molecular biology

The science of the molecular basis of biological activity between biomolecules in the various systems of a cell, including the interactions between DNA, RNA and proteins and their biosynthesis, as well as the regulation of these interactions.

MRI

Magnetic Resonance Imaging—a non-radiation imaging technology using powerful magnets useful in studying the structure of the normal and diseased brain as well as other organs and tissues.

mRNA

Messenger RNA—the RNA produced by a coding gene that leaves the nucleus and migrates to the ribosome to participate in the production of a protein.

Mutation

A permanent alteration of the nucleotide sequence of the genome of an organism. These alterations occur spontaneously during cell replication or as the result of various factors such as radiation, toxins and other factors.

Nanobot

A very small robot or computer device that can be injected into the blood stream or for other uses. They range in size from 1 to 100 nanometers (a nanometer is one billionth of a meter). Potential uses in humans include cancer treatment and brain communication to and from a computer.

Nanotechnology

Manipulation of matter at the atomic, molecular and supermolecular scale including the manufacture of products with at least one dimension in the 1 to 100 nanometer scale (a nanometer is one billionth of a meter).

Natural selection

A key mechanism of evolution in which an organism's traits (phenotype) interacts with the environment to create differential survival and reproduction rates. It is the key concept of Charles Darwin's theory of evolution (Darwinian evolution).

Neanderthal

A member of the extinct species *Homo neanderthalensis* that lived in Eurasia from approximately 600,000 to 40,000 years ago. They are closely related to *Homo sapiens* but not in our direct lineage. There was a small amount of interbreeding between Neanderthals and *Homo sapiens*.

Neuroanatomist

A scientist who studies the anatomy of the brain and other components of the nervous system.

Neuron

The basic cellular unit of the brain and nervous system that transmits, conducts, and receives electrochemical signals from other neurons.

Neuronetwork

An array of neuron connections in a brain and, in artificial intelligence, an array of computer processors simulating the functions of the brain.

Neuroscience

The science of the study of the nervous system.

Neurotechnology

Any technology that allows the study of the brain, visualization of the brain, or repair and treatment of brain disorders.

Neurotransmitter

A chemical substance facilitating the communication between neurons across a synapse. There are many types of endogenous neurotransmitters.

Noncoding gene

A DNA sequence that does not code for a protein but rather codes for various types of RNA important in the epigenome. Thus it is a misnomer to label it as "non-coding." This is distinguished from a "coding gene" (what is usually referred to as a "gene") that codes for a protein.

Nucleotide

An organic molecule that is the building block of DNA and RNA. DNA is composed of four nucleotides: adenine, thymine, cytosine and guanine. In RNA, uracil replaces thymine.

Optogenetics

A technology in which specific neurons are genetically engineered to respond to a specific wavelength of light. The response can be engineered to be either excitatory or inhibitory. This is useful in studying the function of various neuron types and locations. In the study of brain cells, the light is delivered via very thin fiberoptic cables embedded in the brain.

Paleoanthropologist
: A person who studies human evolutionary development and lineages by studying fossils, indirect evidence such as tools and dwellings, and genomics.

Paleoarcheology
: The study of hominid fossils dating to the most recent 15 million years with emphasis on evolution and environmental adaptation.

Paleogenomics
: The study of genomes of extinct species obtained from fossil extractions.

Paleontology
: The study of fossils up to approximately 12,000 years ago.

PCR
: Polymerase Chain Reaction—a molecular biology technique to produce large quantities of DNA from small amounts. Kerry Mullis of the Cetus Corporation received a Nobel Prize for its discovery and development.

PGD
: Pre-implantation Genetic Diagnosis—an IVF technique that allows the complete genetic analysis of a fertilized egg to determine if it is free of specific disease-causing genes or other genetic characteristics prior to being chosen for implantation in a patient.

Phenotype
: The appearance or manifest trait that is determined by a DNA sequence.

Plasmid
: A small DNA molecule within a cell that is physically separated from chromosomal DNA and can replicate independently. They are most commonly found in bacteria as small, circular, double-stranded DNA molecules; however, plasmids are sometimes present in archaea and eukaryotic organisms.

Polymerase Chain Reaction
: A molecular biology technique to produce large quantities of DNA from small amounts. Kerry Mullis of the Cetus Corporation received a Nobel Prize for its discovery and development.

Polypeptide
: A long chain of amino acids forming part or all of a protein. A peptide is a short chain of amino acids.

Postzygotic
: Occurring any time after fertilization and the formation of a zygote including the entire gestational period.

Prezygotic
: Occurring prior to the fertilization and formation of a zygote.

Punctuated equilibrium
A theory in evolutionary biology which proposes that once species appear in the fossil record they will become stable, showing little evolutionary change for most of their geological history. This is in contrast to the more widely held notion that evolutionary changes in species are gradual and continuous over time.

Quadrupedal
Terrestrial locomotion using four legs.

Recombinant DNA
Genetically engineered segments of DNA that are inserted into a genome that would not naturally be found in that genome.

Reflective consciousness
As defined by John Hands in his book, *Cosmosapiens,* the property of an organism by which it is conscious of its own consciousness; that is, not only does it know but also it knows that it knows. John Hands believes that *Homo sapiens* is the only species with this capability.

Regulator
A segment of DNA in a genome that produces RNA or peptides aimed at controlling the expression of a coding gene. It is part of the epigenome.

Reproductive Isolation
The condition in which a species is unable to interbreed with another species to produce viable offspring that are able to have further offspring.

Ribosome
Structures within the cellular cytoplasm in which proteins are produced.

RNA
Ribonucleic Acid—consists of four nucleotides: adenine, uracil, cytosine and guanine.

RNAi
RNA interference—RNA molecules that block the action of a specific messenger RNA thus preventing the protein production of that gene. This is a regulatory mechanism of the epigenome.

rRNA
Ribosomal RNA—a part of the ribosome that interacts with messenger RNA to form the template for protein production.

Savant syndrome
A condition in which a person has extraordinary memory, mathematical, musical and/or artistic ability usually accompanied by autism or other mental disorders.

SC
Species criterion—a characteristic of a species that helps define the species.

Sex chromosome
One of a pair of chromosomes that determines the sex of the individual. In humans, there is one pair of sex chromosomes (X for female, Y for male)

and 22 pairs of autosomal chromosomes. The X chromosome is relatively normal in size and gene count whereas the Y chromosome is short with relatively few coding genes.

Sexual selection
: The mating preference of individuals of one sex of a species for certain physical or behavioral characteristics of the other sex. It is one form of natural selection in evolution in either sympatric or allopatric environments.

Singularitarian
: An individual who promotes the notion that the singularity can be achieved in a relatively short time (less than a century) and can be controlled for the good of humanity.

Singularity
: As used in this book, the point at which human brain intelligence and computer artificial intelligence cannot be distinguished. Ray Kurzweil, Google's director of engineering, popularly promotes this notion.

Somatic cell
: The type of cell that makes up all the tissues of an organism other than the germline cells. Changes in the DNA of somatic cells by mutation or genetic engineering are not passed on to progeny.

Sonogenetics
: A technology in which specific neurons are genetically engineered to respond to a specific sound frequency. The response can be engineered to be either excitatory or inhibitory. This is useful in studying the function of various neuron types and locations.

Speciation
: The process of species origination in evolution.

Species
: The basic unit of taxonomy in the classification of organisms. It typically represents the largest group of organisms that can produce viable offspring that, in turn, are capable of producing further offspring, but are reproductively isolated from all other species.

Species problem
: The difficulty in defining what is a species as reflected in the many different definitions.

Stem cell
: A cell capable of developing into multiple cell types.

Sympatric
: Originating in or occupying the same geographical or physical areas. Contrasted with allopatric.

Synapse
: A small gap between two neurons across which excitation takes place with the participation of neurotransmitter chemicals.

Systematist

A person who studies taxonomy. A taxonomist.

Taxonomy

In biology, the study of the scientific classification of organisms into hierarchical groupings.

TMS

Transcranial Magnetic Stimulation—a technology for non-invasive communication of signals to the brain.

Transgenic

Genetic modification of genes of one species using genes or gene segments from another species.

Trisomy

The abnormal condition where there are three copies of chromosomes of a given type rather than the normal pair. This usually leads to serious clinical conditions.

Vector

In genetic engineering, a biological vehicle used to deliver recombinant DNA into a cell. Plasmids and viruses are commonly used vectors.

Zygote

A eukaryotic cell formed by the union of a male and female gamete (sperm and egg in humans). A zygote has the full complement of DNA to develop into a complete organism.

Index

References

[1] B. Stearns, S. Stearns, *Watching from the Edge of Extinction* (New Haven: Yale University Press, 1999), x.
[2] S. J. Gould, *Wonderful Life* (New York: W. W. Norton & Co., 1989), 24.
[3] E. Kolbert, *The Sixth Extinction*, (New York: Henry Holt & Co., 2014).
[4] Ibid.
[5] G. Ceballos, P. Ehrlich, A. Barnosky, "Accelerated Modern Human-induced Species Loss: Entering the Sixth Mass Extinction." *Science Advances* (June 19, 2015), http://advances.sciencemag.org/content/1/5/e1400253, accessed Aug. 20, 2016.
[6] J. Schopf, in *Ecology of Cyanobacteria II,* B. Whitten, ed. (Dordrecht: Springer Netherlands, 2012).
[7] J. Hands, *Cosmosapiens: Human Evolution from the Origin of the Universe* (New York: Duckworth Publishers, 2016), kindle edition, loc. 11679.
[8] T. Dobzhansky, *Genetics and the Origin of Species* (New York: Columbia University Press, 1937), 4.
[9] S. J. Gould, *Punctuated Equilibrium* (Cambridge: Harvard University Press, 2007).
[10] J. Wilkins, *Species—A History of the Idea* (Berkeley: University of California Press, 2009), kindle edition, loc. 10.
[11] P. Grant, B. Grant, "Hybridization of Bird Species," *Science* 256, no. 5054 (1992):193.
[12] J. Wilkins, *Species—A History of the Idea.*
[13] J. Hey, "The Mind of the Species Problem," *Trends in Ecology and Evolutions* 16, no. 7 (2001): 326.
[14] N. Isaac, J. Mallet, G. Mace, "Taxonomic Inflation: Its Influence on Macroecology and Conservation," *Trends in Ecology and Evolution* 19, no. 9 (2004): 464.
[15] J. Fennessey, T. Bidon, F. Reuss, et. al., "Multi-locus Analyses Reveal Four Giraffe Species Instead of One," *Current Biology*, Sept. 8, 2016, http://www.cell.com/current-biology/fulltext/S0960-9822(16)30787-4, accessed Sept. 9, 2016.
[16] National Human Genome Research Institute, www.genome.gov, accessed Aug. 19, 2016.
[17] E. Mayr, *Systematics and the Origin of Species* (New York: Columbia University Press, 1942).
[18] K. de Queiroz, "Ernst Mayr and the Modern Concept of Species," *Proc Natl Acad Sci USA* 102, Suppl. 1 (May 3, 2005): 6600–6607.
[19] K. de Queiroz, "Species Concepts and Species Delimitation," *Systematic Biology* 56, no. 6 (2007): 879–886.
[20] P. Hebert, A. Cywinska, S. Ball, et al., "Biological Identifications through DNA Barcodes," *Proc Biol Sci* 270, no. 1512 (2003): 313.
[21] R. Ward, R. Hanner, P. Hebert, "The Campaign to DNA Barcode All Fishes," *J. Fish Biol* 74, no. 2 (2009): 329.
[22] A. Clarke, S. Prost, J. Stanton, "From Cheek Swabs to Consensus Sequences: An A to Z Protocol for High Throughput DNA Sequencing of Complete Human Chondrial Genomes," *BMC Genomics* 15 (2014): 68.
[23] R. Cann, "DNA and Human Origins," *Annual Review of Anthropology* 17 (1998): 127.

[24] Integrated Taxonomic Information System, http://www.itis.gov, accessed Aug. 18, 2016.
[25] National Center for Biotechnology Information, www.ncbi.nlm.nih.gov, accessed Aug. 18, 2016.
[26] Kioto Encyclopedia of Genes and Genomes, www.genome.jp/kegg, accessed Aug. 18, 2016.
[27] Encyclopedia of Life, www.eol.org, accessed Aug. 18, 2016.
[28] Species 2000, www.sp2000.org, accessed Aug.18, 2016.
[29] American Ornithologist's Union, http://www.americanornithology.org/content/aou-checklist-north-and-middle-american-birds-7th-edition-and-supplements, accessed Aug. 18, 2016.
[30] American Museum of Natural History, http://research.amnh.org/vz/herpetology/amphibia, Accessed Aug. 18, 2016.
[31] P. Uetz, The Reptile Database, http://reptile-database.org, accessed Aug.18, 2016.
[32] C. Hinchliff, S. Smith, J. Allman, et al., "Synthesis of phylogeny and taxonomy into a comprehensive tree of life." *Proceedings of the National Academy of Sciences* 112 (2015): 12764, https://tree.opentreeoflife.org, accessed Aug. 16, 2016.
[33] C. Mora, D. Tittensor, S. Adl, et al., "How Many Species Are There on Earth and in the Ocean?" *PLoS Biology* (August 23, 2011): DOI: 10.1371/journal.pbio.1001127.
[34] D. Toews, A. Brelsford, C. Grossen, et. al., "Genomic Variation Across the Yellow-rumped Warbler Species Complex," *The Auk* 133, no. 4 (2106):698-717
[35] J. Bond, A. Stockman, *Systematic Biology* 57 (2008): 628–646.
[36] J. Wilkins, *Species—A History of the Idea.*
[37] D. Stamos, *The Species Problem: Biological Species, Ontology, and the Metaphysics of Biology* (New York: Lexington Books, 2003).
[38] J. Brookfield, "Review of *Genes, Categories, and Species* by Jody Hey," *Genet. Res.* 79 (2012): 107.
[39] K. de Queiroz, "Ernst Mayr and the Modern Concept of Species."
[40] *Oxford English Dictionary* (Oxford: Oxford University Press, 2013).
[41] M. Cartwell, F. Smith, *The Human Lineage* (Wiley & Sons, 2011).
[42] D. Schwarz, B. Matta, N. Shakir-Botteri, et al., "Host Shift to an Invasive Plant Triggers Rapid Animal Hybrid Speciation," *Nature* 426, no. 28 (2005): 1038.
[43] Ibid.
[44] J. Coyne, H. Orr, *Speciation* (Sunderland, MA: Sinuaer Associates, Inc., 2004), 30.
[45] C. Darwin, *On the Origin of Species by Means of Natural Selection, or the Preservation of Favored Races in the Struggle for Life* (London: J. Murray, 1859).
[46] J. Rowe, I. Handel, M. Thera, et al., "Blood Group O Protects Against Severe Plasmodium Falciparum Malaria through the Mechanism of Reduced Resetting," *Proc. Nat. Acad. Sci.* 140 (2007): 17471.
[47] R. Dawkins, *The Selfish Gene* (New York, Oxford University Press, 1976)
[48] J. Bull, M. Maron, "How Humans Drive Speciation as well as Extinction," *Proceedings of the Royal Society B* 283 (1833), June 29 2016, DOI: 10.1098/rspb.2016.0600.
[49] L. Trut, American Scientist, http://www.americanscientist.org/issues/pub/early-canid-domestication-the-farm-fox-experiment/1, accessed Aug. 19, 2016.
[50] E. Harris, *Ancestors in Our Genome: The New Science of Human Evolution* (Oxford: Oxford University Press, 2015), 68.
[51] E. Poolman, A. Galvani, "Evaluating Candidate Agents of Selective Pressure for Cystic Fibrosis," *Journal of the Royal Society Interface* published online February 22, 2007, https://doi.org/10.1098/rsif.2006.0154
[52] I. Tattersall, *Masters of the Planet: The Search for Our Human* Origins.
[53] M. Cartmill, F. Smith, *The Human Lineage.*

[54] J. Coyne, H. Orr, *Speciation*, 385.
[55] H. Orr, "The Population Genetics of Speciation: The Evolution of Hybrid Incompatibilities," *Genetics* 139 (1995): 1805.
[56] J. Coyne, H. Orr, *Speciation,* 221.
[57] M. Cartmill, F. Smith, *The Human Lineage.*
[58] D. Johanson, M. Edey, *Lucy: The Beginnings of Humankind* (New York: Simon & Schuster, Inc., 1981) p. 103).
[59] R. Schwartz, "Racial Profiling in Medical Research, *New England Journal of Medicine* 344 (2001):1392-1393
[60] E. Burchard, E. Ziv, N. Coyle, et. al., "The Importance of Race and Ethnic Background in Biomedical Research and Clinical Practice," *New England Journal of Medicine* 348, no. 12 (2003): 1170-1175
[61] A. Lawler, "Mascho Piro Tribe Emerges from Isolation in Peru," *Science* 349, no. 6249 (2015): 679.
[62] R. Fisher, "Has Mendel's Work Been Rediscovered?" *Annals of Science* 1 (1936): 115–137.
[63] R. Fisher, *The Genetical Theory of Natural Selection* (Oxford, Oxford University Press, 1930)
[64] J. Watson, F. Crick, "Molecular Structure of Nucleic Acids," *Nature* 171 (1953): 737.
[65] E. Yong, "The Mysterious Thing About a Marvelous New Synthetic Cell," *Atlantic,* http://www.theatlantic.com/science/archive/2016/03/the-quest-to-make-synthetic-cells-shows-how-little-we-know-about-life/475053, accessed Aug. 19, 2106.
[66] C. Mavragani, I. Sagalovskiy, Q. Guo, et al., "Long interspersed nuclear element-1 retroelements are expressed in patients with systemic autoimmune disease and induce type I interferon," *Arthritis and Rheumatology*, (June 24, 2016), DOI 10.1002/art.39795.
[67] D. Hancks, H. Kazazian, "Active Human Retrotransposons: Variation and Disease," *Current Opinion in Genetics and Development* 22, no. 3 (2012): 191.
[68] ENCODE data describes function of human genome. *NIH News,* Sept. 5, 2012, https://www.genome.gov/27549810/2012-release-encode-data-describes-function-of-human-genome/, accessed Aug. 19, 2016.
[69] M. Dawson, "The Cancer Epigenome: Concepts, Challenges, and Therapeutic Opportunities," *Science* 355, no. 6330 (2017): 1147, DOI 10.1126/science.aam7304
[70] D. D'Onofrio, D. Abel, "Redundancy of the Genetic Code Enables Translational Pausing," *Front. Genet* 5 (2014): 140.
[71] N. Carey, *Junk DNA: A Journey through the Dark Matter of the Genome* (New York: Columbia University Press, 2015).
[72] R. Joehanes, A. Just, R. Marioni, et.al., "Epigenetic Signatures of Cigarette Smoking," *Circulation: Cardiovascular Genetics*, published on-line Sept. 20, 2106, DOI 10.1161/CIRCGENETICS.116.001506, accessed Sept. 21, 2016
[73] I. Donkin, S. Versteyhe, L. Ingerslev, et al., "Obesity and Bariatric Surgery Drive Epigenetic Variation of Spermatozoa in Humans," *Cell Metabolism* 23 (2016): 1, http://dx.doi.org/10.1016/j.cmet.2015.11.004.
[74] L. Hamers, "Why Do Cell's Power Plants Hang on to Their Own Genomes?" *Science* 351, no. 6276 (2016): 903.
[75] M. Ramesh, S. Malik, J. Logsdon, "A Phylogenomic Inventory of Meiotic Genes: Evidence for Sex in Giardia and an Early Eukaryotic Origin of Meiosis," *Curr Biol* 26 (2005): 15.
[76] A. Wilkins, R. Holliday, "The Evolution of Meiosis from Mitosis," *Genetics* 181 (2009): 3.
[77] J. Shendure, J. Akey, "The Origins, Determinants, and Consequences of Human

Mutations," *Science* 349, no. 6255 (2015): 1478.
[78] M. Lodato, M. Woodworth, S. Lee, et al., "Somatic Mutation in Single Human Neurons Tracks Development and Transcriptional History," *Science* 350, no. 6256 (2015): 94.
[79] R. Foley, *Humans before Humanity: An Evolutionary Perspective* (Oxford: Blackwell Publishers, 1995), 110.
[80] L. Wade, DNA from Cave Source Reveals Ancient Human Occupants, *Science* 356, no. 6336 (2017): 363, DOI 10.1126/science.356.6336.363
[81] F. Smith, J. Ahern, *The Origins of Modern Humans: Biology Reconsidered* (Hoboken: Wiley & Sons, 2013) xxi.
[82] S. J. Gould, *Full House* (New York: Harmony Books, 1996), 63.
[83] A. Hill, S. Ward, "Origin of the Hominidae: The Record of African Large Hominoid Evolution Between 14 My and 4 My," *Yearbook of Physical Anthropology* 31 (1988): 49.
[84] D. Wildman, M. Uddin, G. Liu, et al., "Implications of natural selection in shaping 99.4% nonsynonymous DNA identity between humans and chimpanzees: Enlarging genus Homo," *Proc Nat Acad SCI* 100, No. 12 (2003): 7181, DOI: 10.1073/pnas.1232172100, accessed Feb. 12, 2017
[85] E. Harris, *Ancestors in Our Genome.*
[86] Ibid., 33.
[87] Ibid., 92.
[88] D. Gordon, J. Huddleston, M. Chaisson, et al., "Long-read Sequence Assembly of the Gorilla Genome," *Science* 352, no. 6281 (2016): 52, DOI:10.1126/science.aae0344.
[89] E. Harris, *Ancestors in Our Genome,* 114.
[90] Ibid., 128.
[91] I. Tattersall, *The Strange Case of the Rickety Cossack and Other Cautionary Tales from Human Evolution* (St. Martin's Press, 2015).
[92] E. Harris, *Ancestors in Our Genome*, 103.
[93] Y. Kamberov, E. Karisson, G. Kamberova, et al., "A Genetic Basis of Variation in Eccrine Sweat Gland and Hair Follicle Density," *Proc Nat Acad Sci* 112, no. 32 (2015):9932, DOI 10.1073/pnas.1511680112
[94] Mosaic, the Science of Life, http://mosaicscience.com/extra/does-brain-size-matter, accessed Aug. 19, 2016.
[95] S. Herculano-Houzel, "The Remarkable, Yet Not Extraordinary, Human Brain as a Scaled-up Primate Brain and Its Associated Cost," *Proc Nat Acad Sci* 109, Suppl. 1 (2012): 10661.
[96] S. Olkowicz, M. Kocourek, R. Lucan, et al., "Birds Have Primate-like Numbers of Neurons in the Forebrain," http://www.pnas.org/content/early/2016/06/07/1517131113.full.pdf, accessed Aug. 19, 2016.
[97] R. Klein, "Out of Africa and the Evolution of Human Behavior," *Evolutionary Anthropology* 17 (2008): 267.
[98] Y. Harari, *Sapiens: A Brief History of Humankind* (New York: Harper Collins Publishers, 2015), kindle edition, 8.
[99] A. Gibbons, "Why Humans Are the High Energy Apes," *Science* 352, no. 6286(2016): 639.
[100] R. Wrangham, *Catching Fire: How Cooking Made us Human* (New York: Basic Books, 2009), kindle edition, 38, loc. 437.
[101] A. Gibbons, "A New Kind Of Ancestor: *Ardipithecus* Unveiled," *Science* 326, no. 5949 (2009): 36.
[102] T. White, C. Lovejoy, B. Asfaw, et al., "Neither Chimpanzee nor Human, Ardipithecus Reveals the Surprising Ancestry of Both," *PNAS* 112, no. 6 (2014):

4877.
[103] J. Stern, R. Susman, "The Locomotor Anatomy of Australopithecus Afarensis," *Am J Phys Anthropol* 60, no. 3 (1983): 279.
[104] D. Johanson, M. Edey, *Lucy.*
[105] I. Tattersall, *Masters of the Planet: The Search for Our Human Origins* (New York: St. Martin's Press, 2012).
[106] A. Gibbons, "A New Kind Of Ancestor: *Ardipithecus* Unveiled."
[107] K. Wong, "First of Our Kind," in *Becoming Human: Our Past, Present and Future*, published by the editors of *Scientific American* (New York, 2013).
[108] I. Tattersall, *Masters of the Planet: The Search for Our Human* Origins.
[109] D. Johanson, M. Edey, *Lucy.*
[110] C. Lovejoy, "Evolution of Human Walking," *Scientific American* (November 1988), 118.
[111] Personal communication, Donald E. Tyler, PhD, Professor of Biological Anthropology, University of Idaho.
[112] N. Jablonski, G. Chaplin, "The Origin of Hominid Bipedalism Re-examined," *Perspectives in Human Biology 2/Archaeol. Oceania* 27 (1992): 153.
[113] H. McHenry, "Origin of Human Bipedality," *Evolutionary Anthropology* 13 (2004): 116.
[114] P. Wheeler, "The Evolution of Bipedality and Loss of Functional Body Hair in Hominids," *Journal of Human Evolution* 13 (1984): 91, DOI:10.1016/S0047-2484(84)80079-2.
[115] R. Foley, *Humans before Humanity*, 117
[116] P.Rodman, H.McHenry, "Bioenergetics and the Origin of Hominid Bipedalism." Am. J. of Physical Anthropology 52 (1980) 103-106.
[117] Personal communication, April 1, 2017
[118] K. Sayers, C. Lovejoy, "Blood, Bulbs, and Bunodonts: on Evolutionary Ecology and the Diets of Ardipithecus, Australopithecus, and Early Homo," *The Quarterly Review of Biology* 89, no.4 (2014): 1
[119] H. McHenry, "Origin of Human Bipedality."
[120] B. Wood, B. Richmond, "Human Evolution: Taxonomy and Paleobiology," *J. Anat.* 196 (2000): 16.
[121] R. Wrangham, *Catching Fire: How Cooking Made Us Human.*
[122] B. Wood, B. Richmond, "Human Evolution: Taxonomy and Paleobiology."
[123] Institute of Human Origins, Becoming Human, http://www.becominghuman.org/node/homo-heidelbergensis-essay, accessed Aug. 19, 2016.
[124] G. Bräuer, "The Origin of Modern Anatomy: By Speciation or Intraspecific Evolution," *Evolutionary Anthropology* 17 (2008): 23.
[125] A. Gibbons, "Oldest Members of our Species Found in Morocco," *Science* 356, no. 6342 (2017): 993
[126] S. McBrearty, A. Brooks, "The Revolution That Wasn't: A New Interpretation of the Origin of Modern Human Behavior," *Journal of Human Evolution* 39 (2000): 453.
[127] A. Gibbons, "Humanity's Long Lonely Road," *Science* 349, no. 6254 (2015): 1270.
[128] A. Briggs, J. Good, R. Green, et al., "Targeted Retrieval and Analysis of Five Neandertal mtDNA Genomes," *Science* 325 (2009): 318.
[129] M. Krings, A. Stone, R. Schmitz, et al., "Neandertal DNA Sequences and the Origin of Modern Humans," *Cell* 90, no. 1 (1997): 19.
[130] C. Stringer, "The Origin and Evolution of *Homo sapiens,*" *Phil. Trans. R. Soc. B* 371: 20150237, http://dx.doi.org/10.1098/rstb.2015.0237.
[131] D. Papagianni, M. Morse, *The Neanderthals Rediscovered* (London: Thames and Hudson, 2015) kindle edition, loc. 2555.
[132] John Hawks Weblog, http://johnhawks.net/weblog/reviews/neandertals/neandertal_

dna/1000-genomes-introgression-among-populations-2012.html, accessed Aug. 19, 2016.
[133] K. Harmon, "New DNA Analysis Shows Ancient Humans Interbred with Denisovans," *Scientific American* (August 30, 2012), www.scientificamerican.com/article/denisovan-genome, accessed Aug. 19, 2016.
[134] K. Pollard, "What Makes Us Human?" *Scientific American* (May 2009), 44–49.
[135] H. Hu, L. He, K. Fominykh, et al., "Evolution of the Human-specific microRNA miR-491," *Nature Communications* 3 no. 1145 (October 23, 2012).
[136] E. Harris, *Ancestors in Our Genome,* 136.
[137] W. Liu, M. Martinon-Torres, Y. Cai, et al., "The Earliest Unequivocally Modern Humans in Southern China," *Nature* (October 14, 2015), published online, DOI:10.1038/Nature15696.
[138] L. Thompson, "Complete Neanderthal Genome Sequences," *NIH News* (July 11, 2013), www.genome.gov/27539119, accessed Aug. 19, 2016.
[139] M. Raghavan, M. Steinrucken, K. Harris, et al., "Genomic Evidence for the Pleistocene and Recent Population History of Native Americans," *Science* (July 21, 2015).
[140] M. Pedersen, A. Ruter, C. Schweger, et. al., "Postglacial viability and colonization in North America's ice-free corridor," *Nature*, Published online Aug. 10, 2016, doi:10.1038/nature19085.
[141] M. Toups, A. Kitchen, J. Light, "Origin of Clothing Lice Indicates Early Clothing Use by Anatomically Modern Humans in Africa," *Mol. Biol. Evol.* 28, no. 1 (2011): 29.
[142] D. Robson, "When Did Language Evolve," in *The Human Story: New Scientist: The Collection,* Graham Lawton, ed. (London: Reed Business Information Ltd., 2014).
[143] R. Berwick, N. Chomsky, *Why Only Us: Language and Evolution* (Cambridge: MIT Press, 2016).
[144] Y. Harari, *Sapiens,* kindle edition, 4, loc. 19.
[145] C. Stringer, "The Origin and Evolution of *Homo sapiens.*"
[146] K. Prufer, F. Racimo, N. Patterson, "The Complete Genome Sequence of a Neanderthal from the Altai Mountains," *Nature* 505 (2014): 47.
[147] E. Harris, *Ancestors in Our Genome,* p. 175.
[148] C. Stringer, "The Origin and Evolution of *Homo sapiens.*"
[149] A. Gibbons, "Five Matings for Moderns, Neanderthals," *Science* 351, no. 6279 (2016): 1250.
[150] S. Pääbo, *Neanderthal Man: In Search of Lost Genomes* (New York: Basic Books, 2014), kindle edition.).
[151] S. Pääbo, "The Diverse Origins of the Human Gene Pool," *Nature Reviews/Genetics* 16 (2015): 313.
[152] S. Pääbo, *Neanderthal Man,* loc. 4527.
[153] E. Harris, *Ancestors in Our Genome,* 183.
[154] S. Condemi, A. Mounier, P. Giunti, et al. "Possible Interbreeding in Late Italian Neanderthals? New Data from the Mezzena Jaw (Monti Lessini Verona, Italy)," *PLOS One* 10 (2013): 1371.
[155] C. Duarte, J. Mauricio, P. Pettitt, et al., "The Early Upper Paleolithic Human Skeleton from the Abrigo do Lagar Velho (Portugal) and Modern Human Emergence in Iberia," *PNAS* 96, no. 13 (1999): 1073.
[156] M. Currat, L. Excoffier, "Strong Reproduction Isolation Between Humans and Neanderthals Observed from Inferred Patterns of Introgression," *Proc Nat Acad Sci* 108, no. 37 (2011): 15129.
[157] D. Papagianni, M. Morse, *The Neanderthals Rediscovered,* loc. 971.
[158] E. Harris, *Ancestors in Our Genome,* 184.

[159] M. Marshall, "The Others," in *The Human Story: New Scientist: The Collection,* Graham Lawton, ed. (London: Reed Business Information Ltd., 2014).
[160] C. Simonti, B. Vernot, L. Bastarache, et al., "The Phenotypic Legacy Admixture between Modern Humans and Neandertals." *Science* 351, no. 6274 (2016): 737.
[161] E. Culotta, "A Single Wave of Migration from Africa Peopled the Globe," *Science* 354, no. 6319 (2016): 1522
[162] S. Lopez, L. van Dorp, G. Hellenthal, "Human Dispersal Out of Africa: A Lasting Debate," *Evol. Bioinform. Online* 11, suppl 2 (2105): 57, DOI 10.4137/EBO. S33489
[163] W. Haak, L. Lazaridis, N. Patterson, "Massive Migration from the Steppe was a Source for Indo-European Languages in Europe," *Nature* 522 (2015): 207.
[164] C. Stringer, "The Origin and Evolution of *Homo sapiens.*"
[165] A. Gibbons, "Prehistoric Eurasians Streamed into Africa, Genome Shows," *Science* 350, no. 6257 (2015): 149, DOI: 10.1126/science.350.6257.149.
[166] J. Viegas, Seeker, http://news.discovery.com/human/evolution/neanderthal-skeleton-provides-evidence-of-interbreeding-with-humans-130327.htm, accessed Aug. 20, 2016.
[167] Y. Harari, *Sapiens.*
[168] M. Sisk, J. Shea, "The African Origin of Complex Projectile Technology: An Analysis Using Tip Cross-sectional Area and Perimeter," *International Journal of Evolutionary Biology* 2011 (2011): 968012, DOI 10:4061/2011/968012.
[169] C. Houldcroft, S. Underdown, "Neanderthal Genomics Suggests a Pleistocene Time Frame for the First Epidemiologic Transition," *Am J of Physical Anthropology* (April 10, 2016), DOI 10.1002/ajpa.22985.
[170] P. Shipman, *The Invaders: How Human and Their Dogs Drove Neanderthals to Extinction* (The Belknap Press of Harvard University Press, 2015).
[171] A. Nowell, "All Work and No Play Makes a Dull Child," in *The Human Story: New Scientist: The Collection,* Graham Lawton, ed. (London: Reed Business Information Ltd., 2014).
[172] F. Welker, M. Hajdinjak, S. Talamo, et. al., "Palaeoproteomic evidence identifies archaic hominins associated with the Châtelperronian at the Grotte du Renne," *PNAS* September, 2016, online DOI 10.1073/pnas.1605834113, accessed September 23, 2016
[173] R. Foley, *Humans before Humanity*, 214
[174] E. Kolbert, *The Sixth Extinction.*
[175] S. Gould, *Full House*, 175.
[176] J. Rosen, "Thinking the Unthinkable," *Science* 353, no. 6296 (2016): 232.
[177] E. Kolbert, *The Sixth Extinction.*
[178] D. Robertson, M. Mckenna, O. Toon, et al., "Survival in the First Hours of the Cenozoic," *GSA Bulletin* 116 (2004): 760.
[179] D. Denkenberger, J. Pearce, *Feeding Everyone No Matter What: Managing Food Security After Global Catastrophe*, (London: Elsevier, 2015).
[180] P. Ward, J. Kirschvink, *A New History of Life: The Radical New Discoveries About the Origins and Evolution of Life on Earth* (New York: Bloomsbury Press, 2015).
[181] E. Botkin-Kowacki, "Did That Asteroid Have Help Wiping Out the Dinosaurs?" *Christian Science Monitor* (October 1, 2015), http://www.csmonitor.com/Science/2015/1001/That-asteroid-might-have-needed-help-wiping-out-the-dinosaurs, accessed Aug. 20, 2016.
[182] "Toba Catastrophe Theory," Wikipedia, https://en.wikipedia.org/wiki/Toba_catastrophe_theory, accessed Aug. 20, 2016.
[183] P. Ward, J. Kirschvink, *A New History of Life.*
[184] J. Hawks, E. Wang, G. Cochran, et al., "The Recent Acceleration of Human

Adaptive Evolution," *PNAS* 104, no. 52 (2007): 20753.

[185] C. Stringer, "The Origin and Evolution of *Homo sapiens*."

[186] Ibid.

[187] S. Fan, M. Hansen, Y. Lo, et. al., "Going Global by Adapting Local: A Review of Recent Human Adaptation." *Science* 354 no. 6308 (2016): 54

[188] F. Cocks, "Global Warming vs. the Next Ice Age," *MIT Technology Review*, https://www.technologyreview.com/s/416786/global-warming-vs-the-next-ice-age/, accessed Aug. 20, 2016.

[189] S. Prusiner, "Cell Biology: A Unifying Role for Prions in Neurodegenerative Diseases," *Science* 336, no. 6088 (2012): 1511.

[190] Y. Harari, *Sapiens*.

[191] J. Zalasiewicz, "What Mark Will We Leave on the Planet," *Scientific American*, Sept. 2016, http://www.scientificamerican.com/index.cfm/_api/render/file/?method=attachment&fileID=DFB9E60B-A9B2-4373-A9D8952819D01B04, accessed Sept. 4, 2016

[192] C. Stringer, "The Origin and Evolution of *Homo sapiens*."

[193] P. Thompson, T. Cannon, K. Narr, "Genetic Influences on Brain Structure," *Nature Neuroscience* 4, no. 12 (2001): 1253.

[194] The White House, https://www.whitehouse.gov/share/brain-initiative, accessed Aug. 20, 2016.

[194] Human Brain Project, https://www.humanbrainproject.eu, accessed Aug. 20, 2016.

[196] G. Marcus, J. Freeman, eds., *The Future of the Brain* (Princeton: Princeton University Press, 2015), kindle edition, 208, loc 3521.

[197] E. Underwood, "The Brain's Identity Crisis," *Science* 349 no. 6248 (2015): 575.

[198] M. Kaku, *The Future of the Mind* (New York: Doubleday, 2014).

[199] B. Libet, C. Gleason, E. Wright, et al., "Time Of Conscious Intention to Act in Relation to Onset of Cerebral Activity (Readiness-Potential): The Unconscious Initiation of a Freely Voluntary Act," *Brain* 106 (1983): 623.

[200] Y. Harari, *Homo Deus*: A Brief History of Tomorrow, (London: Harvill Secker, 2016): 109

[201] J. Hands, *Cosmosapiens*, loc. 9445.

[202] M. Tomasello, M. Carpenter, Shared Intentionality, *Developmental Science* 10, no.1 (2007): 121

[203] S. Sloman, P. Fernbach, *The Knowledge Illusion: Why We Never Think Alone* (New York: Riverhead Books, 2017)

[204] R. Dunbar, Neocortex Size as a Constraint on Group Size in Primates, *Journal of Human Evolution* 20 (1992): 469

[205] C. Sagan, *The Dragons of Eden: Speculations on the Evolution of Human Intelligence* (Ballantine Publishing Co., 1977), kindle edition, 7, loc. 169.

[206] National Institute of Neurological Disorders and Stroke, http://www.ninds.nih.gov/disorders/brain_basics/genes_at_work.htm, accessed Aug. 20, 2016.

[207] W. Mischel, Y. Shoda, M. Rodriguez, "Delay of Gratification in Children," *Science* 244, no. 4907 (1989): 933.

[208] B. Casey, L. Somerville, I. Gotlib, et al., "Behavior and Neural Correlates of Delay of Gratification 40 Years Later," *PNAS* 108, no. 36 (2011): 14988.

[209] Guinness World Records, http://www.guinnessworldrecords.com/world-records/most-pi-places-memorised, accessed Aug. 20, 2016.

[210] D. Tammet, *Born on a Blue Day* (New York: Free Press, 2006), 2.

[211] A. LePort, A. Mattfield, H. Dickinson-Anson, "Behavioral and Neuroanatomical Investigation of Highly Superior Autobiographical Memory (HSAM)," *Neurobiology of Memory and Learning* 98, no. 1 (2012): 78.

[212] A. Stocco, C. Prat, D. Losey, et al., "Playing 20 Questions with the Mind:

Collaborative Problem Solving by Humans Using a Brain to Brain Interface," *PLoS ONE* 10, no. 9: e0137303, http://journals.plos.org/plosone/article?id=10.1371/journal.pone.0137303, accessed Aug. 20, 2016.
[213] J. O'Shea, V. Walsh, "Transcranial Magnetic Stimulation," *Current Biology* 17, no. 6 (2007): R196.
[214] M. Cerauskaite, Health Units, http://www.healthunits.com/scientists-create-first-brain-to-text-system, accessed Aug. 20, 2016.
[215] M. Kaku, *The Future of the Mind.*
[216] C. King, P. Wang, C. McCrimmon, et al., "The Feasibility of a Brain-computer Interface Functional Electrical Stimulation System for the Restoration of Overground Walking after Paraplegia," *Journal of NeuroEngineering and Rehabilitation* 12 (2015): 80.
[217] A. Ajiboye, F. Willett, D. Young, et al., "Restoration of reaching and grasping movements through brain-controlled muscle stimulation in a person with tetraplegia: a proof-of-concept demonstration," *Lancet,* Online First, March 28, 2017, DOI: 10.1016/S0140-6736(17)30601-3
[218] Emotiv, www.emotiv.com, accessed Aug. 20 2016.
[219] Nicolelis Lab, http://www.nicolelislab.net, accessed Aug. 20, 2016.
[220] M. Pais-Vieira, G. Chiuffa, M. Lebedev, "Building an Organic Computing Device with Multiple Interconnected Brains," *Scientific Reports* 5 (July 9, 2015). http://www.nature.com/srep/2015/150706/srep11869/pdf/srep11869.pdf., accessed Aug. 20, 2016.
[221] M. Sparkes, *The Telegraph*, http://www.telegraph.co.uk/technology/10567942/Supercomputer-models-one-second-of-human-brain-activity.html, accessed Aug. 20, 2016.
[222] C. Koch, "Project Mindscope," in *The Future of the Brain*, 25, loc. 532.
[223] M. Kaku, *The Future of the Mind.*
[224] Ibid.
[225] R.Fisher, *The Genetical Theory of Natural Selection*
[226] G. Meisenberg, "The Reproduction of Intelligence," *Intelligence* 38 (2010): 220.
[227] Headlines and Global News, http://www.hngn.com/articles/118117/20150811/world-population-11-billion-2100-due-growth-africa.htm, accessed Aug. 20, 2016.
[228] S. Aridi, Christian Science Monitor, http://www.csmonitor.com/Environment/2015/0814/Resource-overdraft-Planet-Earth-crosses-into-ecological-red, accessed Aug. 20, 2016.
[229] M. Farah, "The Unknowns of Cognitive Enhancement," *Science* 350, no. 6259 (2015): 379, DOI: 10.1126/science.aad5893.
[230] D. Jackson, R. Symons, P. Berg, "Biochemical Method for Inserting New Genetic Information into DNA of Simion Virus 40: Circular SV40 DNA Molecules Containing Lambda Phage Genes and the Galactose Operon of *Escherichia Coli,*" *PNAS* 69, no. 10 (1972): 2904.
[231] S. Cohen, A. Chang, L. Hsu, "Nonchromosomal Antibiotic Resistance in Bacteria: Transformation of *Escherichia coli* by R-factor DNA," *PNAS* 69, no. 8 (1972): 2110.
[232] S. Cohen, A. Chang, H. Boyer, et al., "Construction of Biologically Functional Bacterial Plasmids *in vitro,*" *PNAS* 70, no. 11 (1973): 3240.
[233] A. Chang, S. Cohen, "Genome Construction between Bacterial Species *in vitro*: Replication and Expression of *Staphylococcus* Plasmid Genes in *Escherichia coli,*" *PNAS* 71, no. 4 (1974): 1030.
[234] D. Nicholl, *An Introduction to Genetic Engineering, Third Edition* (Cambridge University Press, 2008).
[234] R. Jaenisch, B. Mintz, "Simian Virus 40 DNA Sequences in DNA of Health Adult

Mice Derived from Preimplantation Blastocysts Injected with Viral DNA," *PNAS* 71, no. 4 (1974): 1250.
[236] S. Rojahn, MIT Technology Review, http://www.technologyreview.com/view/518171/glow-in-the-dark-rabbits, accessed Aug. 20, 2016.
[237] Wikipedia, https://en.wikipedia.org/wiki/Asilomar_Conference_on_Recombinant_DNA, accessed Aug. 20, 2016.
[238] S.Mukherjee, *The Gene: An Intimate History* (New York: Scribner, 2016).
[239] "Online Mendelian Inheritance in Man," McKusick-Nathans Institute of Genetic Medicine, Johns Hopkins University, http://omim.org, accessed Aug. 20, 2016.
[240] J. Bartlett, D. Stirling, "A Short History of the Polymerase Chain Reaction," *Methods in Molecular Biology* 226 (2003): 3.
[241] J. Heier, S. Kherani, S. Desai, et. al., Intravitreous Injection of AAV2-sFLT01 in Patients with Advanced Neovascular Age-related Macular Degeneration: a Phase 1 Open-label Trial, *Lancet online,* May 16, 2017, http://dx.doi.org/10.1016/s0140-6736(17)30979-0
[242] K. Culver, W. Anderson, R. Blaese, "Lymphocyte Gene Therapy," *Human Gene Therapy* 2, no. 2 (1991): 107.
[243] U. Dave, N. Jenkins, N. Copeland, "Gene Therapy Insertional Mutagenesis Insights," *Science* 303 (2004): 333.
[244] S. De Ravin, X. Wu, N. Theobald, et al., "Lentiviral Hematopoietic Stem Cell Gene Therapy for Older Patients with X-linked Severe Combined Immunodeficiency," abstract presented at the 57th Annual Meeting of the American Society of Hematology, December 6, 2015.
[245] O. Obasogie, "Ten Years Later: Jesse Gelsinger's Death and Human Subjects Protection," Bioethics Forum, October 22, 2009, Center for Genetics and Society, http://www.geneticsandsociety.org/article.php?id=4955, accessed Aug. 20, 2016.
[246] S. Boseley, "Gene Therapy Treatment for Cystic Fibrosis may be Possible by 2020, Scientists Say," *Guardian* (July 3, 2015), http://www.theguardian.com/science/2015/jul/03/gene-therapy-cystic-fibrosis-2020-scientists, accessed Aug. 20, 2016.
[247] J. Ribeil, S. Hacein-Bey-Abina, E. Payen, et al., "Gene Therapy in a Patient with Sickle Cell Disease," *NEJM* 376, (2017): 848
[248] J. Wilson, "Glybera's Story Mirrors That of Gene Therapy," *GEN* (January 1, 2013), http://www.genengnews.com/gen-articles/glybera-s-story-mirrors-that-of-gene-therapy/4671, accessed Aug. 20, 2016.
[249] K. Boztug, M. Schmidt, A. Schwarzer, et al, "Stem-cell Gene Therapy for the Wiskott-Aldrich Syndrome," *NEJM* 363, no. 20 (2010): 1918.
[250] A. Wierzbicki, A. Viljoen, "Alipogene Tiparvovec: Gene Therapy for Lipoprotein Lipase Deficiency," *Expert Opinion on Biological Therapy* 13, no. 1 (2013): 7, DOI: 10.1517/14712598.2013.738663.
[251] M. Wadman, Antisense Rescues Babies from Killer Disease, *Science* 354, no. 6318 (2016):1359, http://www.sciencemagazinedigital.org/sciencemagazine/16_december_2016?sub_id=BOV1xNikfGyBo&u1=41278342&folio=1359&pg=11#pg11, accessed Dec. 16, 2016
[252] National Cancer Institute, http://www.cancer.gov/about-cancer/treatment/research/car-t-cells, accessed Aug. 20, 2016.
[253] Blood Cancer Treatment Called "Revolutionary" after All Study Patients Responded, CBS news, June 5, 2017, http://www.cbsnews.com/news/blood-cancer-multiple-myeloma-gene-therapy-all-study-patients-responded/, accessed June 5, 2017
[254] B. Teh, H. Ishiyama, W. Mai, et al., "Long-term outcome of a phase II trial using immunomodulatory in situ gene therapy in combination with intensity-modulated radiotherapy with or without hormonal therapy in the treatment of prostate cancer,"

Journal of Radiation Oncology (December 12, 1915), published online: http://link.springer.com/article/10.1007%2Fs13566-015-0239-y, accessed Aug. 20, 2016.
[255] R. Andorno, "The Oviedo Convention: A European Legal Framework at the Intersection of Human Rights and Health Law," *Journal of International Business and Law* 2 (2005): 133.
[256] American Medical Association, http://www.ama-assn.org/ama/pub/physician-resources/medical-science/genetics-molecular-medicine/current-topics/gene-therapy.page, accessed Aug. 20, 2016.
[257] M. Araki, T. Ishii, "International Regulatory Landscape and Integration of Corrective Genome Editing into *In Vitro* Fertilization," *Reproductive Biology and Endocrinology* 12 (2014): 108, DOI:10.1186/1477-7827-12-108.
[258] *Mitochondrial Replacement Techniques: Ethical, Social and Policy Considerations* (Washington, DC: National Academies Press, 2016).
[259] H. Greely, *The End of Sex and the Future of Human Reproduction* (Cambridge: Harvard University Press, 2016).
[260] M. Jinek, K. Chylinski, I. Fonfara, et al., "A Programmable Dual-RNA-guided DNA endonuclease in Adaptive Bacterial Immunity," *Science* 337, no. 6069 (2012): 816.
[261] E. Pennisi, "The CRISPR Craze," *Science* 341, no. 6148 (2013): 833.
[262] P. Knoepfler, *GMO Sapiens: The Life-Changing Science of Designer Babies* (New Jersey: World Scientific Publishing Co., 2016).
[263] S. Zhang, Wired, "The War Over Genome Editing Just Got a Lot More Interesting," http://www.wired.com/2015/09/war-genome-editing-just-got-lot-interesting, accessed Aug. 20, 2016.
[264] E. Pennisi, "The CRISPR Craze."
[265] H. Ledford, "The Unsung Heroes of CRISPR," *Nature* 535 no. 7612 (2016): 342, doi:10.1038/535342a.
[266] V. Gantz, N. Jasinskiene, O. Tatarenkova, et al., "Highly Efficient Cas9-mediated Gene Drive for Population Modification of the Malaria Vector Mosquito *Anopheles stephensi,*" http://www.pnas.org/content/112/49/E6736.full?sid=943dcdcf-6a7e-48b6-bb5c-c56266526aa0, accessed Aug. 20, 2016.
[267] C. Larson, "China's Bold Push into Genetically Customized Animals," *Scientific American* (November 17, 2015), http://www.scientificamerican.com/article/china-s-bold-push-into-genetically-customized-animals, accessed Aug. 20, 2016.
[268] D. Cyranoski, *Nature*, http://www.nature.com/news/chinese-scientists-to-pioneer-first-human-crispr-trial-1.20302?WT.mc_id=TWT_NatureNews, accessed Aug. 20, 2016.
[269] K. Schaefer, W. Wu, D. Colgan, et al., Unexpected Mutations after CRISPR-Cas9 Editing *in vivo*, Correspondence, *Nature Methods* 14, no. 6 (2017): 547.
[270] F. Zhang, Y. Wen, X. Guo, "CRISPR/Cas9 for Genome Editing: Progress, Implications and challenges," *Human Molecular Genetics* 23, R1 (2014): R40.
[271] P. Liang, Y. Xu, X. Zhang, et al., "CRISPR-Cas9-mediated Gene Editing in Human Tripronuclear Zygotes," *Protein & Cell* 6, no. 5 (2015): 363.5.
[272] D. Cryanoski, S. Reardon, "Embryo Editing Sparks Epic Debate," *Nature* 520 (2015): 593.
[273] Radiolab, http://www.radiolab.org/story/antibodies-part-1-crispr, accessed Aug. 20, 2016.
[274] R. Stein, "Swedish Scientist Starts DNA Experiments On Healthy Human Embryos," http://www.npr.org/2016/09/22/495069492/swedish-scientist-starts-dna-experiments-on-healthy-human-embryos, accessed September 23, 2016
[275] D. Gibson, J. Glass, C. Lartigue, et al., "Creation of a Bacterial Cell by a Chemically Synthesized Genome," *Science* 329, no. 5987 (2010): 52, DOI: 10.1126/

science.1190719.
[276] S. Richardson, L. Mitchell, G. Stracquadanio, et al., "Design of a Synthetic Yeast Genome," *Science* 355, no. 6327 (2017): 1040
[277] L. Shekhtman, The Christian Science Monitor, http://www.csmonitor.com/Science/2016/0516/Why-did-Harvard-scientists-hold-a-secret-synthetic-genome-meeting, accessed Aug. 20, 2016.
[278] D. Cyranoski, "Gene-edited Micropigs to Be Sold as Pets at Chinese Institute," *Nature* 526 (October 1, 2015):18, DOI:10.1038/nature.2015.18448.
[279] M. Piper, L. Partridge, D. Raubenheimer, et al., "Dietary Restriction and Aging: A Unifying Perspective," *Cell Metabolism* 14, no. 2 (2011): 154.
[280] S. Leiser, H. Miller, R. Rossner, et al., "Cell Nonautonomous Activation of Flavin-containing Monooxygenase Promotes Longevity and Gealth Span," *Science* 350, no. 6266 (2015): 1375, DOI: 10.1126/science.aac9257.
[281] E. Blackburn, E. Epel, *The Telomere Effect,* Grand Central Publishing, New York, 2017
[282] http://bioviva-science.com/blog/first-gene-therapy-successful-against-human-aging-2, accessed 3/18/17.
[283] A. Ocampo, P. Reddy, P. Martinez-Redondo, et. al., "In Vivo Amelioration of Age-Associated Hallmarks by Partial Reprogramming," *Cell* 167, no.7 (2016): 1719
[284] Human Aging Genomics Resources, http://genomics.senescence.info/genes/stats.php, accessed Aug. 20, 2016.
[285] L. Cassidy, M. Narita, "GATA Get a Hold on Senescence," *Science* 349, no. 6255 (2015): 1448.
[286] SENS Research Foundation, www.sens.org, accessed Aug. 20, 2016.
[287] A. de Grey, *The Singularity and the Methuselarity: Similarities and Differences, in Strategy for the Future of Health*, R. G. Bushko, ed. (IOS Press, 2009).
[288] D. Koshland, "The Seven Pillars of Life," *Science* 295, no. 5563 (2002): 2215.
[289] S. Reardon, New Life for Pig-to-Human Transplants, *Nature* 527 (2015): 152, DOI: 10.1038/527152a
[290] B. Zetsche, J. Gootenberg, O. Abudayyey, et al., "Cpf1 Is a Single RNA-guided Endonuclease of a Class 2 CRISPR Cas System," *Cell* (September 25, 2015), http://www.cell.com/cell/abstract/S0092-8674(15)01200-3, accessed Aug. 20, 2016.
[291] I. Slaymaker, L. Gao, B. Zetsche, et al., "Rationally Engineered Cas9 Nucleases with Improved Specificity," *Science* 351, no. 6268 (2015): 84, DOI:10.1126/science.aad5227.
[292] B. Rauch, M. Silvis, J. Hultquist, et. al., "Inhibition of CRISPR-Cas9 with Bacteriophage Proteins", *Cell* (Dec. 29, 2016), DOI: 10.1016/j.cell.2016.12.009, accessed 12/30/16
[293] A. Komor, Y. Kim, M. Packer, et al., "Programmable Editing of a Target Base in Genomic DNA without Double-Stranded DNA Cleavage," *Nature* (April 20, 2016), DOI:10.1038/nature17946.
[294] L. Silver, *Remaking Eden: How Genetic Engineering and Cloning Will Transform the American Family* (New York: Avon Books, 1998).
[295] N. Wade, *Before the Dawn: Recovering the Lost History of Our Ancestors,* (New York, Penguin Press, 2006), kindle edition, 277.
[296] B. Joy, "Why the Future Doesn't Need Us," *Wired Magazine* (April 2000).
[297] P. Knoepfler, *GMO Sapiens.*
[298] J. Savulescu, N. Bostrom, eds., *Human Enhancement* (Oxford: Oxford University, 2009).
[299] J. Hands, *Cosmosapiens.*
[300] D. Koshland, "The Seven Pillars of Life."
[301] J. White, E. Southgate, J. Thomson, et al., "The Structure of the Nervous System

of the Nematode Caenorhabditis elegans," *Phil. Trans. R. Soc. London B*, 314, no.1165 (1986): 1. DOI: 10.1098/rstb.1986.0056, accessed Jan. 19, 2017

[302] S. Emmons, "The Beginning of Connectomics: A Commentary on White et al. (1986), 'The Structure of the Nervous System of the Nematode Caenorhabditis elegans,'" *Phil. Trans. R. Soc. London B Biol Sci* 370, no. 1666 (2015): 20140309, DOI: 10.1098/rstb.2014.0309.

[303] S. Hill, "Whole Brain Simulation," in *The Future of the Brain,* 113, loc. 1927.

[304] International Neuroinformatics Coordinating Facility, www.incf.org, accessed Aug. 20, 2016.

[305] M. Ahrens, M. Orger, D. Robson, et al., "Whole-brain Functional Imaging at Cellular Resolution Using Light-sheet Microscopy," *Nature Methods* 10 (2013): 413.

[306] Allen Brain Atlas, http://help.brain-map.org/display/mouseconnectivity/ALLEN+Mouse+Brain+Connectivity+Atlas, accessed Aug. 20, 2016.

[307] Human Connectome Project, http://www.humanconnectomeproject.org, accessed Aug. 20. 2016.

[308] M. Hawrylycz, "Building Atlases of the Brain," in *The Future of the Brain*, 8, loc. 259.

[309] M. Glasser, T. Coalson, E. Robinson, et al,, "A Multi-modal Parcellation of Human Cerebral Cortex," *Nature*, 2016, doi:10.1038/nature18933.

[310] A. Hodges, *Alan Turing: The Enigma* (Princeton: Princeton University Press, 1983), 316, loc. 6445.

[311] D. Ferrucci, J. E. Brown, Chu-Carroll, R. et. al., "Building Watson: An Overview of the DeepQA Project'" *AI Magazine,* 2010, http://www.aaai.org/Magazine/Watson/watson.php, accessed Oct. 8, 2016.

[312] M. Blois, "Clinical Judgment and Computers," *NEJM* 303 (1980): 192.

[313] K. Naughton, "Humans Are Slamming into Driverless Cars Exposing a Key Flaw," *Automotive News* (December 18, 2015), https://www.autonews.com/article/20151218/OEM11/151219874/humans-are-slamming-into-driverless-cars-and-exposing-a-key-flaw, accessed Aug. 20, 2016.

[314] Google, DeepMind, www.deepmind.com, accessed Aug. 20, 2016.

[315] M. Moraveik, M. Schmid, N. Burch, et al., DeepStack: Expert-level Artificial Intelligence in Heads-up No-Limit Poker, *Science* 356, no. 6337 (2017): 508

[316] R. Kurzweil, *The Singularity Is Near* (New York: Viking, 2005).

[317] M. Waldrop, "Computer Modelling: Brain in a Box," *Nature* 482, no. 7386 (2012): 456, DOI:10.1038/482456a.

[318] Ibid., 9, loc 383.

[319] http://www.cnbc.com/2017/03/27/elon-musk-is-founding-another-company.html, accessed 3/28/17

[320] https://www.youtube.com/watch?v=ZrGPuUQsDjo, accessed 3/18/17

[321] H. Huang, M. Sakar, A. Petruska, et al., "Soft micromachines with programmable motility and morphology," *Nature Communications* 7 no.12263 (2016), doi:10.1038/ncomms12263.

[322] J. Ho, A. Yeh, E. Neofytou, "Wireless Power Transfer to Deep Tissue Microimplants," *PNAS* 111, no. 22 (2014): 7974.

[323] K. Montgomery, A. Yeh, J. Ho, et al., "Wirelessly Powered, Fully Internal Optogenetics for Brain, Spinal and Peripheral Circuits in Mice," *Nature Methods* 12 (2015): 969.

[324] C. Newmarker, "Eric Topol on How to Prevent Heart Attacks with Nanosensors," http://www.qmed.com/mpmn/medtechpulse/eric-topol-how-prevent-heart-attacks-nanosensors, accessed Aug. 20, 2016.

[325] Q. Hu, W. Sun, C. Qian, et al., "Anticancer Platelet-mimicking Nanovehicles. Advanced Materials," published online September 29, 2015, DOI: 10.1002/

adma.201503323.

[326] R. Xu, G. Zhang, J. Mai, et al., "An Injectable Nanoparticle Generator Enhances Delivery of Cancer Therapeutics," *Nature Biotechnology* (March 14, 2016), published online: DOI 10.1038/nbt.3506.

[327] S. Douglas, I. Bachelet, G. Church, "A Logic-gated Nanorobot for Targeted Transport of Molecular Payloads," *Science* 335, no. 6070 (2012): 831.

[328] A. von Eschenbach, "Elimination the Suffering and Death due to Cancer by 2015," Manhattan Institute for Policy Research, Medical Progress Bulletin No. 1 (September 2005).

[329] T. Blackwell, "Is the War on Cancer an 'Utter Failure'? A Sobering Look at How Billions in Research Money Is Spent," *National Post* (March 15, 2013).

[330] H. Markram, E. Muller, S. Ramaswamy, et al., "Reconstruction and Simulation of Neocortical Microcircuitry," *Cell* 163, no. 2 (2015): 456.

[331] K. Kupferschmidt, "Virtual Rat Brain Fails to Impress Its Critics," *Science* 350, no. 6258 (2015): 263, http://www.sciencemagazinedigital.org/sciencemagazine/16_october_2015?sub_id=BOV1xNikfGyBo&folio=263&pg=15#pg15.

[332] Open Message to the European Commission, http://neurofuture.eu, accessed Aug. 20, 2016.

[333] M. Enserink, "A 1 Billion Brain Reboot," *Science* 347, no. 6229 (2015): 1406, DOI: 10.1126/science.347.6229.1406.

[334] A. Moore, NIH lecture, Recent Developments in Artificial Intelligence—Lessons from the Private Sector (September 21, 2015), given by Webinar.

[335] M. Carandini, "From Circuits to Behavior: A Bridge Too Far?" in *The Future of the Brain*, 177, loc. 3003.

[336] R. Kurzweil, *How to Create a Mind: The Secret of Human Thought Revealed* (New York: Viking, 2012), kindle edition, 9, loc.232.

[337] L. Chang, D. Tsao, The Code for Facial Identity in the Primate Brain, *Cell* 169, no. 6 (2017): 1013, DOI http://dx.doi.org/10.1016/j.cell.2017.05.011, accessed June 6, 2017

[338] A. Pascual-Leone, "The Brain That Plays Music and Is Changed by It," *Annals of the New York Academy of Science* 930 (2001): 315.

[339] C. Koch, G. Marcus, "Neuroscience in 2064—A Look at the Last Century as Told to Christof Koch and Gary Marcus," in *The Future of the Brain*, 255, loc. 4208.

[340] J. Renne, The Immortal Ambitions of Ray Kurzweil: A Review of *Transcendent Man*, *Scientific American*, February 15, 2011, https://www.scientificamerican.com/article/the-immortal-ambitions-of-ray-kurzweil/

[341] G. Ross, "An Interview with Douglas R. Hofstadter," *American Scientist Online* (January 2007), http://www.americanscientist.org/bookshelf/pub/douglas-r-hofstadter, accessed Aug. 20, 2016.

[342] Singularity University, http://singularityu.org/overview, accessed Aug. 20, 2016

[343] E. Underwood, "Lifelong Memory May Reside in Nets Around Brain Cells," *Science* 350, no. 6260 (2015): 491, http://www.sciencemagazinedigital.org/sciencemagazine/30_october_2015?sub_id=BOV1xNikfGyBo&folio=491&pg=11#pg11.

[344] P. Allen, M. Greaves, "The Singularity Isn't Near," *MIT Technology Review* (October 12, 2011), http://www.technologyreview.com/view/425733/paul-allen-the-singularity-isnt-near, accessed Aug. 20, 2016.

[345] M. Ford, *Rise of the Robots: Technology and the Threat of a Jobless Future* (New York: Basic Books, 2015).

[346] BBC News interview by Rory Cellan-Jones of Stephen Hawking, December 2, 2014, http://www.bbc.com/news/technology-30299992, accessed Aug. 20, 2016.

[347] L. Grossman, "2045: The Year Man Becomes Immortal," *Time Magazine* (February

10, 2011).

[348] V. Vinge, "VISION-21 Symposium sponsored by NASA Lewis Research Center and the Ohio Aerospace Institute, March 30–31, 1993," http://mindstalk.net/vinge/vinge-sing.html, accessed Aug. 20, 2016.

[349] M. Prigg, Daily Mail, http://www.dailymail.co.uk/sciencetech/article-2548355/Google-sets-artificial-intelligence-ethics-board-curb-rise-robots.html, accessed Aug. 20, 2016.

[350] S. Hawking, M. Tegmark, S. Russell, "Transcending Complacency on Superintelligent Machines," *Huffington Post* (April 19, 2014), http://www.huffingtonpost.com/stephen-hawking/artificial-intelligence_b_5174265.html, accessed Aug. 20, 2016.

[351] N. Bostrom, *Superintelligence: Paths, Dangers, Strategies* (Oxford: Oxford University Press, 2014).

[352] Ibid., loc. 6029

[253] J. Barrat, *Our Final Invention: Artificial Intelligence and the End of the Human Era* (New York: Thomas Dunne Books, 2013).

[354] M. Shanahan, "The Technological Singularity," *The MIT Press Essential Knowledge Series* (Cambridge: MIT Press, 2015).

[355] F. Heylighen, "Return to Eden? Promises and Perils on the Road to Global Superintelligence," in *The End of the Beginning: Life, Society and Economy on the Brink of the Singularity,* Ben and Ted Goertzel, eds. (Humanity+ Press, 2015).

[356] M. Anissimov, *Our Accelerating Future: How Superintelligence, Nanotechnology, and Transhumanism Will Transform the Planet* (Berkeley: Zenit Books, 2015).

[357] Ibid.

[358] OpenAI, https://openai.com/blog/introducing-openai, accessed Aug. 20, 2016.

[359] Machine Intelligence Research Institute, https://intelligence.org, accessed Aug. 20, 2016.

[360] Foresight Institute, www.foresight.org, accessed Aug. 20, 2016.

[361] Future of Life Institute, http://futureoflife.org, accessed Aug. 20, 2016.

[362] Future of Humanity Institute, https://www.fhi.ox.ac.uk, accessed Aug. 20, 2016.

[363] Humanity +, www.humanityplus.org, accessed Aug. 20, 2016.

[364] I. Good, "Speculations Concerning the First Ultraintelligent Machine," in Franz L. Alt and Morris Rubinoff, *Advances in Computers, Volume 6* (New York: Academic Press, Inc., 1965).

[365] V. Vinge, "VISION-21 Symposium sponsored by NASA Lewis Research Center and the Ohio Aerospace Institute, March 30–31, 1993."

[366] J. Markoff, *Machines of Loving Grace: The Quest for Common Ground between Humans and Robots* (New York, Harper Collins, 2015), kindle edition, loc. 591.

[367] R. Kurzweil, *The Singularity Is Near,* 209.

[368] N. Agar, "Whereto Transhumanism? The Literature Reaches a Critical Mass," Hastings Center Report (May–June 2007).

[369] G. McDonald, Seeker, http://news.discovery.com/tech/robotics/hitchhiking-robot-embarks-on-u-s-vacation-150720.htm, accessed Aug. 20, 2015.

[370] A. Shademan, R. Decker, J. Opfermann, et al., "Supervised Autonomous Robotic Soft Tissue Surgery," *Science Translational Medicine* 8, no. 337 (2016): 337, DOI: 10.1126/scitranslmed.aad9398.

[371] J. Markoff, *Machines of Loving Grace.*

[372] D. Alba, "It's Your Fault Microsoft's Teen AI Turned into Such a Jerk," *Wired Online* (March 25, 2016), http://www.wired.com/2016/03/fault-microsofts-teen-ai-turned-jerk, accessed, Aug. 20, 2016.

[373] J. Kaplan, *Humans Need Not Apply: A Guide to Wealth and Work in the Age of Artificial Intelligence* (New Haven: Yale University Press, 2015).

[374] M. Ford, *Rise of the Robots: Technology and the Threat of a Jobless Future* (New York: Basic Books, 2015).
[375] M. Brain, *The Second Intelligent Species: How Humans Will Become as Irrelevant as Cockroaches* (Raleigh: BYG Publishing, Inc., 2015).
[376] J. Kaplan, *Humans Need Not Apply.*
[377] J. Hands, *Cosmosapiens.*
[378] S. Gould, *Full House*, 214
[379] J. Hamzelou, "Exclusive: World's first baby born with new "3 parent" technique." *New Scientist*, published online Sept. 27, 2016, https://www.newscientist.com/article/2107219-exclusive-worlds-first-baby-born-with-new-3-parent-technique/, accessed September 28, 2016.
[380] J. Coyne, H. Orr, *Speciation*, 10.
[381] K. de Queiroz, personal communication.
[382] J. Coyne, H. Orr, *Speciation*, 15.
[383] R. May, P. Harvey, "Species Uncertainties," *Science* 323, no. 5915 (2009): 687.
[384] J. Padial, A.,Miralles, I. De La Riva, et al., "The Integrative Future of Taxonomy," *Frontiers in Zoology* 7 (2010): 16.
[385] C. Stringer, *Lone Survivor* (New York: St. Martin's Press, 2012).
[386] C. Darwin, *The Next Million Years* (London: Rupert Hart-Davis, 1952).
[387] G. Radick, "Beyond the 'Mendel-Fisher' Controversy." *Science* 350, no. 6257 (2015): 159, DOI: 10:1126/science.aab3846.
[388] Hemophilia-information.com, Homecare for the Cure, http://www.hemophilia-information.com/hemophilia-genetics.html, accessed Aug. 20, 2016.
[389] J. Hang, R. Wan, C. Yan, et al., "Structural Basis of Pre-mRNA Splicing," *Science* 349, no. 6253 (2015): 1191.
[390] N. Liscovitch-Brauer, S. Alon, H. Porath, et al., "Trade-off between Transcriptome Plasticity and Genome Evolution in Cephalopods," *Cell* 169, no. 2 (2017):191, DOI: 10.1016/j.cell.2017.03.025
[391] B. Sullenger, S. Nair, "From the RNA World to the Clinic," *Science* 352, no. 6292 (2016): 1417, DOI:10:1126/science.aad8709.
[392] D. D'Onofrio, D. Abel, "Redundancy of the Genetic Code Enables Translational Pausing."
[393] E. Pennisi, "Circular DNA Throws Biologists for a Loop." *Science* 356, no. 6342 (2017): 996
[394] S. Dimauro, G. Davidzon, "Mitochondrial DNA and Disease," *Annals of Medicine* 37 (2005): 222.
[395] R. Lightowlers, R. Taylor, D. Turnbill, "Mutations Causing Mitochondrial Disease: What Is New and What Challenges Remain," *Science* 349, no. 6255 (2015): 1494.
[396] D. Pritchard, "The Girl with Three Biological Parents," http://www.bbc.com/news/magazine-28986843, accessed Aug. 20, 2016.

CPSIA information can be obtained
at www.ICGtesting.com
Printed in the USA
LVHW04*1701250418
574832LV00009B/161/P